Genomics of Plant–Pathogen Interaction and the Stress Response

Plants are an indispensable part of human and animal lives for nutrition and health. But pests, diseases and abiotic stress adversely affect crop yield, which ultimately places significant pressure on society to provide food to an increasing population. Moreover, it also encourages increased chemical/pesticide usage on crops, which we see in the biomagnification of toxic and hazardous compounds polluting water bodies, soil and the environment. This condition will continue to worsen in the future due to the resistance-acquiring ability of pathogens against plant defense and chemical treatments. In addition, environmental disturbances and consumer health issues are being reported more promptly than before due to intensive use of pesticides in food production. Plant diseases affect our daily lives, as the use of insecticides and pesticides has become part of our food chain. As a result, precise disease diagnosis and management is crucial in order to avoid huge losses in plant production and related commodities. Accurate detection, precise diagnosis and proper management can play a significant role in keeping plants free from pathogens. In this book, scientists, researchers and scholars share their research knowledge, offering a valuable resource for understanding plant diseases, pathogen interaction and responses to stress through an omics perspective, contributing to further advancements in the field.

Diseases in plants may be caused by various factors, such as viruses, bacteria, fungi and abiotic stress. Disease causes low crop yield, production of poor-quality fruits and grains, and deficiency of nutrients, which have a direct impact on human and animal health. A genomics-based approach can be applied to disease diagnosis; disease outbreak; evolution of plant and pathogen genome for disease outbreak in relation to climate change; and development of long-term strategies for plant health and defense. This book presents an overview of omics technologies and approaches used to understand:

1) the relation between plants and their environment in terms of diseases
2) responses to abiotic stress
3) the genomics of plant–pathogen interaction
4) herbicide-resistance mechanisms
5) the epigenetics of plant–pathogen interaction
6) gene regulation during abiotic stress response
7) the oxidative stress response

Genomics of Plant–Pathogen Interaction and the Stress Response

Edited by
Ashutosh Mani and Sandeep Kushwaha

CRC Press
Taylor & Francis Group
Boca Raton London New York

CRC Press is an imprint of the
Taylor & Francis Group, an **informa** business

First edition published 2024
by CRC Press
6000 Broken Sound Parkway NW, Suite 300, Boca Raton, FL 33487–2742

and by CRC Press
4 Park Square, Milton Park, Abingdon, Oxon, OX14 4RN

CRC Press is an imprint of Taylor & Francis Group, LLC

Library of Congress Cataloging-in-Publication Data
Names: Mani, Ashutosh, editor. | Kushwaha, Sandeep, editor.
Title: Genomics of plant pathogen interaction and the stress response / edited by Ashutosh Mani and Sandeep Kushwaha.
Description: First edition | Boca Raton, FL : CRC Press, 2024 | Includes bibliographical references and index.
Identifiers: LCCN 2023016896 (print) | LCCN 2023016897 (ebook) | ISBN 9780367721084 (hardback) | ISBN 9780367721091 (paperback) | ISBN 9781003153481 (ebook)
Subjects: LCSH: Plant-pathogen relationships. | Genomics. | Plants—Effect of stress on—Genetic aspects.
Classification: LCC SB732.7 .G46 2024 (print) | LCC SB732.7 (ebook) | DDC 632/.3—dc23/eng/20230628
LC record available at https://lccn.loc.gov/2023016896
LC ebook record available at https://lccn.loc.gov/2023016897

ISBN: 978-0-367-72108-4 (hbk)
ISBN: 978-0-367-72109-1 (pbk)
ISBN: 978-1-003-15348-1 (ebk)

DOI: 10.1201/9781003153481

Typeset in Times
by Apex CoVantage, LLC

Contents

Editor Biographies

Dr Ashutosh Mani holds an MSc and a PhD in Bioinformatics from the University of Allahabad, India. He worked as a MIUR fellow at the University of Cagliari, Italy, and is currently an assistant professor at Motilal Nehru National Institute of Technology Allahabad, India. He has more than ten years of teaching and research experience in molecular and structural biology, genomics and computational biology. His areas of interest include herbal drug discovery against Alzheimer's disease and diabetes and plant stress tolerance. Dr Mani has published more than 50 articles in national and international journals of repute. He has supervised 5 PhD, 10 MTech and 23 BTech students. He has completed four sponsored research projects and has received research grants from funding agencies such as DST, SERB, ICMR and DBT.

Dr Sandeep Kushwaha started his research career as a doctoral student at the Department of Bioinformatics, Maulana Azad National Institute of Technology, Bhopal. During his stay at Lund University and the Swedish University of Agricultural Sciences, he worked on more than ten major scientific projects in plant and soil science, elucidating molecular and genetic mechanisms and plant interaction using omics data and bioinformatics tools and techniques. He works as Scientist-E at the National Institute of Animal Biotechnology, Hyderabad, India. He has completed various research projects, published research articles in the public domain and taught various higher education courses in India as well as abroad.

Contributors

Iffat Zareen Ahmad
Natural Products Laboratory, Department of Bioengineering, Integral University, Dasauli, Kursi Road, Lucknow, India

Supriya Babasaheb Aglawe
ICAR – National Research Center on Grapes, Manjari Farm, Pune, India

Amrita Chakraborty
Faculty of Forestry and Wood Sciences, Czech University of Life Sciences Prague, Czech Republic

Selvi Chellamuthu
Department of Agricultural Entomology, Tamil Nadu Agricultural University, Coimbatore, India

Yamini Deepthi
Crop improvement section, Indian Institute of Rice Research, Hyderabad, India

Sonam Dwivedi
Natural Products Laboratory, Department of Bioengineering, Integral University, Dasauli, Kursi Road, Lucknow, India

R. K. Gaur
Department of Biotechnology, D. D. U. Gorakhpur University, Gorakhpur, Uttar Pradesh, India

Sellamuthu Gothandapani
Faculty of Forestry and Wood Sciences, Czech University of Life Sciences Prague, Czech Republic

Manish Kumar Gupta
Department of Biotechnology, Veer Bahadur Singh Purvanchal University, Jaunpur, India

Dhandapani Gurusamy
Department of Botany, Kongunadu Arts and Science College (Autonomous), Bharathiar University, Coimbatore, India

Kateřina Hamouzová
Department of Agroecology and Crop Production, Faculty of Agrobiology, Food and Natural Resources, Czech University of Life Sciences Prague, Prague, Czech Republic

Pramod Katara
Centre of Bioinformatics, University of Allahabad, Prayagraj, Uttar Pradesh, India

Elhan Khan
Natural Products Laboratory, Department of Bioengineering, Integral University, Dasauli, Kursi Road, Lucknow, India

Himanshu Kishore Prasad
Department of Life Science and Bioinformatics, Assam University Silchar, Assam, India

Pavlína Košnarová
Department of Agroecology and Crop Production, Faculty of Agrobiology, Food and Natural Resources, Czech University of Life Sciences Prague, Prague, Czech Republic

Pragati Kumari
Scientist Hostel-S-02, Chauras campus, Srinagar Garhwal, Uttarakhand, India

Sandeep Kushwaha
National Institute of Animal Biotechnology, Hyderabad, India

Ashutosh Mani
Department of Biotechnology, Motilal Nehru National Institute of Technology Allahabad, Uttar Pradesh, India

Dibyalochan Mohanty
Department of Pharmaceutics, Anurag University, Hyderabad, India

Shrishti Naidu
Department of Botany, Hansraj College, University of Delhi, India

Chitra Nehra
Department of Biosciences, Mody University of Sciences and Technology, Lakshmangarh, Rajasthan, India

S. J. S. Rama Devi
School of Life Sciences, Sambalpur University, Burla, Sambalpur, India

Sakthi Ambothi Rathnasamy
Centre for Plant Molecular Biology and Biotechnology, Tamil Nadu Agricultural University, Coimbatore, India

Amit Roy
Excelentní tým pro mitigaci (ETM), Faculty of Forestry and Wood Sciences, Czech University of Life Sciences, Prague, Czech Republic

Haram Sarfraz
Natural Products Laboratory, Department of Bioengineering, Integral University, Dasauli, Kursi Road, Lucknow, India

Madhab Kumar Sen
Department of Agroecology and Crop Production, Faculty of Agrobiology, Food and Natural Resources, Czech University of Life Sciences Prague, Prague, Czech Republic

Archana Singh
Department of Botany, Hansraj College, University of Delhi, Delhi, India

Bhupender Singh
School of Bioengineering and Biosciences, Lovely Professional University, Punjab, India

Indrakant K Singh
Molecular Biology Research Lab, Department of Zoology, Deshbandhu College, University of Delhi, Kalkaji, New Delhi, India

Josef Soukup
Department of Agroecology and Crop Production, Faculty of Agrobiology, Food and Natural Resources, Czech University of Life Sciences Prague, Prague, Czech Republic

Saurav Suman
Department of Botany, Hansraj College, University of Delhi, Delhi, India

Rahul Thakur
Department of Biotechnology, Hemvati Nandan Bahuguna Garhwal (Central) University, Srinagar Garhwal, Uttarakhand, India

Aneeta Tiwari
Natural Products Laboratory, Department of Bioengineering, Integral University, Dasauli, Kursi Road, Lucknow, India

Shivani Tyagi
Centre of Bioinformatics, University of Allahabad, Prayagraj, Uttar Pradesh, India

Atul Kumar Upadhyay
Department of Biotechnology, Thapar Institute of Engineering & Technology Patiala, India

Birendra Singh Yadav
Centre of Bioinformatics, University of Allahabad, Allahabad, India

Dinesh Yadav
Department of Biotechnology, D.D.U. Gorakhpur University, Gorakhpur, India

Saurabh Yadav
Department of Biotechnology, Hemvati Nandan Bahuguna Garhwal (Central) University, Srinagar Garhwal, Uttarakhand, India

1 Integrative Omics
Trends and Scope for Agriculture

Pramod Katara, Shivani Tyagi, and Manish Kumar Gupta

1.1 INTRODUCTION

Omics can be defined as a collection of various technologies, especially high throughput technologies, in nature (e.g., NGS, microarray, mass spectrometry, NMR), used to understand and examine the role, relationships and actions of the genome (genomics), transcripts (transcriptomics), proteins (proteomics) and metabolites (metabolomics), along with other essential molecules involved in the governance and maintenance of various biological and cellular processes. Since its beginning, scientists have exhaustively utilized omics for different aspects of various human studies, but in recent decades a number of agriculture studies have also been reported as utilizing omics to improve agriculture, to some extent, in different ways. The major benefited areas are predictability in plant breeding, production of stress-resistant crops, production of disease-resistant crops, production of crops with high nutritional values and drought-resistance new crop varieties in water-limiting conditions (Álvarez-Venegas & De-la-Peña, 2016). While there are many omics, the major omics, which are crucial in central dogma and responsible for molecular information flow from genotype to phenotype, are genomics, transcriptomics, proteomics and metabolomics.

1.1.1 Genomics

Genomics deals with the genome, which harbors complete molecular information about an individual and decides its fate. To understand the physiology of any plant, it is very important to understand its molecular information, which resides in the genome. Along with human genome sequencing, scientists have also undertaken sequencing of the plant genome; for instance, *Arabidopsis thaliana* marked the beginning of the plant genomics era. After completing the genome of *Arabidopsis thaliana* within a short period of time, scientists reported in early 2000 the genome sequence of the first edible rice crop. The availability of genome information provides opportunities to use knowledge-based approaches for various agriculture purposes (Dixit et al., 2019). The major processes supported by genomics are plant biotechnology, molecular breeding and marker-assisted selection, which speed up the development of new crop varieties (Van Emon, 2016).

DOI: 10.1201/9781003153481-1

1.1.2 Transcriptomics

The transcription process, through the expression of the genes, produces a transcript; a complete set of these transcripts is known as a transcriptome and provides information to cell machinery to produce proteins. Transcriptomics provides a complete study and understanding of these transcriptomes. It provides expression profiling of a cell at particular given conditions that summarize the expression levels of mRNAs in a given cell population. A transcriptome totally depends on the requirements of a particular cell at a particular time period/under particular environmental conditions; thus, it remains dynamic in nature. Transcriptomes help to assess the effect of a given stress (biotic/abiotic) on the cell and the expression of the gene of importance, thus helping to define the role of genes in plant physiology or, sometimes, patho-physiology (Singh et al., 2016). In earlier days, transcriptomics was mainly based on expressed sequence tags, but the development of various high throughput technologies, i.e., cDNA microarray, RNA-seq, has sped up the process and enhanced the capability for further understanding of functional genomics and plant physiology (Jaiswal et al., 2012).

1.1.3 Proteomics

Proteins are the main functional molecules of the cell and present ubiquitously in plants. Transcripts transfer the gene information into functional molecules, as they are not able to perform this function themselves. Proteins can be considered as the product of the translation process that shows expression on transcripts. Like transcriptomes, proteomes are also dynamic in nature; constituting the proteome depends on the condition of the respective cell. Almost all of the biological functions of the cell are performed by the proteins, which also play key roles for the texture, yield, flavor and nutritional value of virtually all plant-food products. As mentioned, a proteome is directly affected by the surrounding environment of the cell, which is mainly influenced by its biological (infection, disease) and environmental stimuli (e.g., stress, temperature, drought). Proteomics helps agriculture scientists to get insight into protein function and their associations with disease and other features related to their utilities. Techniques like X-ray crystallography and NMR provide structural information about the proteins, which help scientists to understand the structural component of importance. The development of bioinformatics offers a huge range of computational software/tools that provide the ability to perform proteomics-related analysis that helps with understanding various mechanisms of agriculture importance, i.e., disease/resistance, pesticide functionality, identification of nutritious protein components, etc.

1.1.4 Metabolomics

Metabolomics is the youngest branch of omics and deals with the large-scale study of metabolome, the total content of metabolite within the cell or tissue under consideration. In short, metabolomics deals with the study of small molecules (50–1500 daltons) that formed as a product of metabolism and their interaction among themselves within a given biological system. Metabolomics is crucial as the study of these small molecules provides a link between genotypes and phenotypes, as well as a picture of how these metabolites are affected by genetic and environmental factors. The metabolome is more dynamic than transcriptomes and proteomes, as it is subject to internal as well as environmental conditions (Van Emon, 2016). Despite its complex nature, metabolomics is more powerful than other omics in terms of gaining insight into molecular-phenotype and providing direct connections between biochemical activity and cell/tissue conditions. Mass spectrometry (MS) and nuclear magnetic resonance (NMR) techniques facilitate the use of metabolic profiling and metabolic fingerprinting. Use of metabolic profiling and metabolic fingerprinting offers a way to monitor the effect of biotic/abiotic stress on cell metabolism and phenotype and are already used in practice for association studies between metabolism and phenotype. These profiles provide useful information for agriculture scientists and are helpful in the development of improved crop varieties; at the same time, they enhance the basic understanding of plant systems biology. Use of metabolomics in agriculture provides the scope for (1) testing of the effects of various biotic and abiotic stresses on plant health/food quality, (2) quality improvements in nutrition and plant health, (3) increasing nutritive values, and (4) distinguishing the mode of action of pesticides, providing critical information for new pesticide discoveries (Van Emon, 2016; Gomez-Casati et al., 2013).

1.1.5 Other Important Omics for Agriculture

Metagenomics: Metagenomics is a comparatively new branch of genomics that deals with microbial studies. It provides a way to analyze microbial DNA directly acquired from environmental samples, i.e., agriculture fields, sewage water, etc. As metagenomics does not require isolation, purification and culture of the sample, it provides scope for genomic analysis of unculturable microbes as well as a way to research the microbial community present in a particular environment. Soil microbes play a crucial role in various agriculture-related aspects (nitrogen fixation, host–pathogen interactions, microbial disease, etc.), and metagenomics opens the door to perform research on them for better agriculture outputs.

Epigenomics: Epigenomics deals with the complete set of reversible modifications on the genetic material of a cell (without changing DNA sequence), known as the epigenome.

By altering the accessibility of regulatory complexes to the genome, epigenetic changes (DNA methylation, histone posttranslational modifications) regulate gene expression.

These changes affect a number of crucial agronomically significant properties, including blooming time, fruit development, plant immunology, and responses to environmental conditions. Although very few studies have examined the role of epigenetic mechanisms in agricultural research, it is anticipated that they will soon be used, mostly to boost crop output in difficult situations (Álvarez-Venegas & De-la-Peña, 2016).

Fluxomics: Fluxomics can be considered as a branch of metabolomics, and it deals with approaches to determining the fluxome, or rate of metabolic reaction in a biological entity of interest. As fluxomics directly defines metabolic phenotypes with response to the surrounding environment, it is considered an important tool for observing the effect of various biotic/abiotic stresses on plants. One frequently used technique in fluxomics is metabolic flux analysis (MFA), which relies on isotopically or radioactively labeled substrate.

Phenomics: Phenomics deals with complete set of physical and biochemical traits, known as phenome. It defines phenotypes that can be produced by a given organism during the development process, as well as in response to intrinsic and extrinsic stress. Rapid development in high-throughput phenotyping technologies allows agriculture scientists to perform in-field measurements of biologically relevant phenotypes that provide a range of scope for agriculture (Rexroad et al., 2019; Zhao et al., 2019)

Interactomics: Interactomics deals with interactomes and provides insight about the interaction between any small and large macromolecules in a given cell with particular reference. It offers inclusive interaction information at molecular, cellular and biological levels, e.g., gene–gene interaction, gene-regulatory network, protein–protein interaction network (PPIN).

1.2 BIOINFORMATICS AS THE BACKBONE OF OMICS

All omics utilize high-throughput technologies and, as a product, produce a bulk amount of data that must be stored, processed and analyzed to retrieve meaning. Bioinformatics offers different data management systems, algorithms and software to fulfill these requirements (Katara, 2014). In this era of high-throughput omics, it is almost impossible to imagine any research without the use of bioinformatics, and agriculture is no exception. Currently, hundreds of databases are available for every layer of omics. Initially, these resources were more focused towards the human, but since agriculture research has begun to expand, sped-up agriculture bioinformatics has also grown in an exponential manner. Bioinformatics provides a range of approaches and resources (Yadav et al., 2020) that can be utilized by agriculture scientists to handle agriculture data-related aspects (Tables 1.1 and 1.2).

1.3 INTEGRATIVE OMICS: OMICS TO MULTI-OMICS

Each layer of omics has the potential to deal with problems on its own, and provides a comprehensive evaluation of molecular phenotypes (Pinu et al., 2019), but as each of the omics is confined within its own layer, they are unable to provide cross-talk between layers, which is required to understand the complete systemic picture of the system. To understand any biological system completely, it is essential to consider all of its elements together (Yadav et al., 2020).

TABLE 1.1
Omics Data Sources and Major Computational Resources of Agriculture Importance

Omics	Data Source	Major Database Resources	Software/Tools
Genomics	Sequencing technology, i.e., NGS (WGS, WES)	ENA, SRA, Phytozome, Plantgdb, MaizeGDB, Gramene	Blast, T-coffe, Galaxy, MEGA
Transcriptomics	cDNA- microarray, ESTs, SAGE, RNA-Seq	GEO, SRA, dbESTs, UniGene, Bgee, SwissRegulon	Cluster, Arabidopsis Co-expression Tool (ACT), ESTscan, ESTAnnotator, ISMARA, QuasR
RNomics	Microarray	Rfam, miRbase, MirGeneDB, MGI, NONCODE, PDBe, TAIR, snoDB, lncRNAdb, LncBase	R2DT, RNAhybrid, Diana-microT, micro Tar, miTarget, Target Scan, mfold, vsfold, RNAfold, Sfold, RNAstructure, UNAfold, SPOT-RNA
Proteomics	Protein microarray, X-ray crystallography, MR, MS, homology modeling	PDB PIR, Swiss-Prot, STRING.	Modeller, Swiss-modeler, OpenMS
Glycomics	MS, HPLC, multiple reaction monitoring, Glycan microarray	CSDB, Glyco3D, Glyconnect, GlyGen, MatrixDB, UniCarbDB, UniCarbKB, UniLectin, Glycostore, CFG Database, GlyTouCan,	DrawGlycan, GLYCAM-Web, Glycan Analyzer, GLAD, GlycoMod GlycoGlyph, DANGO, Glydin, GlycoSiteAlign
Metabolomics	NMR, mass spectrometry (GC-MS, LC-MS, CE-MS)	MetaboLights, MetaCyc, METLIN, MassBank, METAGENE, MetaboLights, Reactome, Plant Metabolic Network (PMN)	MetaboloAnalyst, MeltDB, BioSpider, XCMS Online
Metagenomics	NGS	ENA, SRA	MetaWRAP, MEGAN-LR
Phenomics	Multi-platform remote sensing technologies, 3D imaging, MRI	Phenomicsdb	Deep Plant, phenomics, henopsis, phenoscope
Fluxomics	Metabolic flux analysis	CeCaFDB	COBRA toolbox, 13CFLUX2
Interactomics	X-ray crystallography, affinity chromatography, Y2H	STRING, Intact	Cell Designer, Cytoscape, Gephi

TABLE 1.2
Major Objective of Omics with Related Bioinformatics Approaches

Omics	Major Objective	Bioinformatics Approaches
Genomics	• Function determination of genes and the elements • Sequence variation and polymorphism detection • Develop and apply genome-based strategies for the early detection, diagnosis, and treatment of disease • Develop new technologies to study genes and DNA on a large scale and store genomic data efficiently	• Sequence homology • Sequence polymorphism • Molecular phylogeny • Gene network • Novel gene finding or identification
Transcriptomics	• To catalogue all species of transcript • To determine the transcriptional structure of genes, in terms of their start sites, 5′ and 3′ ends, splicing patterns and other posttranscriptional modifications • To quantify the changing expression levels of each transcript during development and under different conditions	• Genome wide gene expression analysis • Splicing variants • Differential expression analysis
Proteomics	• Protein function prediction • Protein activity observations • To identify the protein interaction, i.e., protein–protein interaction, DNA–protein interaction, protein–drug interactions	• Structure modeling • Active site prediction • Protein–protein interaction, • Protein dynamics (structure to function) analysis
Metabolomics	• Establish the connection between genotype and phenotype • Identification of metabolome in given condition, i.e., disease, stress, infection • Identification of new disease indicators (markers) • Observation of metabolic reactions	• Metabolic network • Metabolic pathways • Metabolic profiling/fingerprinting • Flux analysis

Systems Biology: Systems biology is a method that examines all of an entity's components collectively to provide a broad understanding of the system, such as an organism, tissue, or cell. Systems biology assumes that all elements of any system are always interlinked and governed by each other (Figure 1.1). It is not

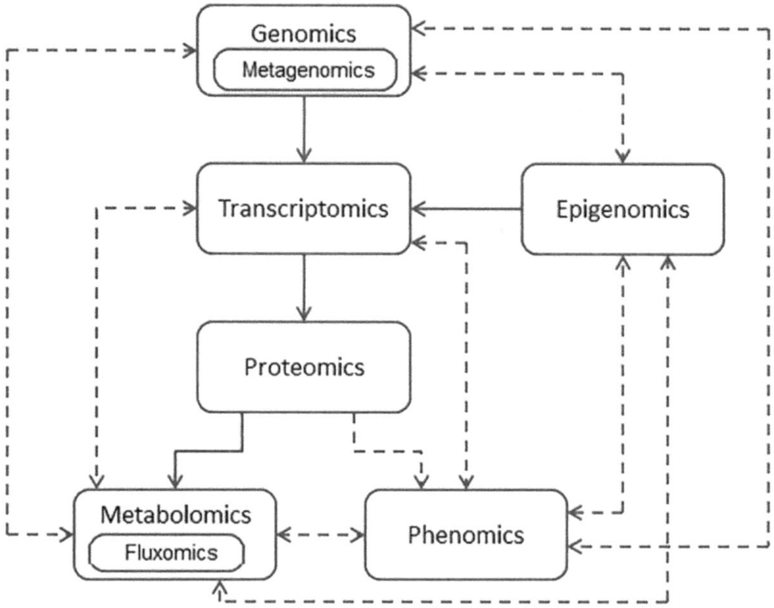

FIGURE 1.1 Direction and connectivity/dependencies of different omics.

possible to understand them independently, and thus a reductionist approach cannot provide complete and reliable insights into any system.

Systems biology integrates numerous bio-molecular data with well-designed statistical models to unravel the biology of complex living systems. It does this by drawing on interdisciplinary elements (primarily biology, chemistry, statistics, physics, and engineering).

The objective of systems biology raises the obvious need to combine the data generated from multiple omics approaches to create a more holistic understanding of systems as related to their growth, adaptation, development and progression in response to stress (Pinu et al., 2019).

1.4 CURRENT TRENDS IN INTEGRATIVE OMICS

The bottleneck of all key multi omics studies has been the integration of data. A number of scientists and experts from various backgrounds are needed in this process for input and interpretation. To assess the quality and efficacy of an experiment design, together with the qualitative data obtained from this instrument, it is necessary to have a few such experts. For the analysis and interpretation of multiomics data, the following three main approaches can be used (Pinu et al., 2019). These techniques include the following: (1) data integration after analysis, (2) integrated data analysis and (3) system modelling (Pinu et al., 2019).

1.4.1 Post-Analysis Data Integration Approaches

Currently, post-analysis data integration approaches are widely used for integrative analysis. In this approach, different omics data sets are first analyzed individually, and then, based on the studies; their key features are integrated together and finally produce a significant conclusion (Figure 1.2). Though these kinds of conclusions are multi-omics–based, due to several bottlenecks, such as different scales of omics data, they mostly compromise with coverage of the complete data. At the same time, this approach relies on humans for various purposes,

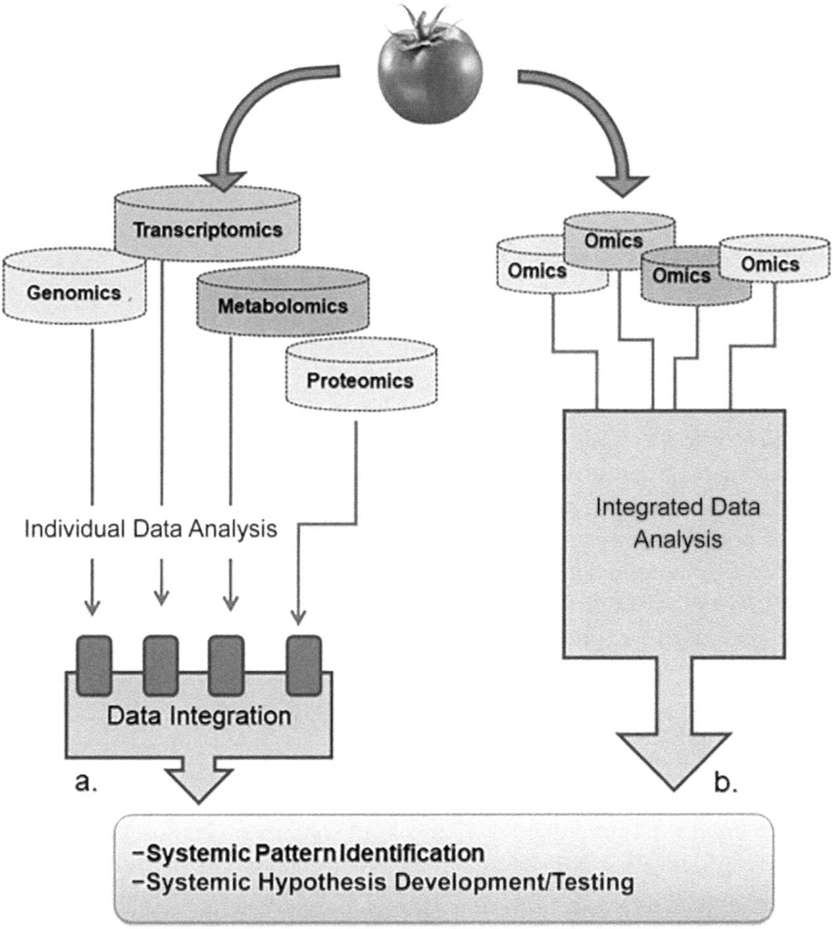

FIGURE 1.2 Integration of multi-omics (a) post-analysis data integration approaches (b) integrated data analysis approaches.

e.g., selection of key features for integration, data interpretation, etc., and is thus sensitive to human biases.

1.4.2 Integrated Data Analysis Approaches

An integrated approach to Data Analysis uses various specialized software tools that merge different sets of omics data prior to doing any further analysis and interpretation (Figure 1.2). Using the integrated pipelines that allow automatic data integration and analysis so as to reduce the possibility of bias in humans, it is therefore possible to statistically generate common similarities between each omics approach and platform (Pinu et al., 2019). Due to its integrative workflow, it is more reliable for multi-omics analysis, but as it lacks gold standard and purpose-specific pipelines, it is at present not preferred by agriculture scientists.

1.4.3 Systems Modeling

The integrative approaches mentioned thus far deal with real observation-based omics data for a particular system in a given condition; most of the time, these approaches face practical problems with regard to data integration and conclusion. Under such conditions, systems modeling and simulation techniques can be used for the studies of complex biosystems. In order to compare new experimental findings with modeled predictions, integration methods relying on a good defined understanding of the system under investigation are used. Such understanding is often based on comprehensive, pre-existing genetics, transcriptomics and metabolomics data for the system to be studied (Pinu et al., 2019). As systems modeling rely on pre-existing models, it cannot be applied to new systems.

1.5 BIOINFORMATICS RESOURCES FOR THE PRACTICE OF INTEGRATIVE OMICS

As a result of the advancement of systems biology, integrative omics needs triggered bioinformatics to cope with the requirements of dealing with multi-omics data, and bioinformatics responded well, allowing multi-omics data to be handled in an integrative manner, so that it can be easily accessed for cross-layer analysis purposes. The most ideal example of these types of integration is NCBI (National Centre for Biotechnology Information, www.ncbi.nlm.nih.gov/), which provides a link between and across layers of data; for instance, Genbank data (genomics) is connected to a GEO (transcriptomics) and protein database (proteomics). Bioinformatics also provides software and tools with multi-omics data analysis pipelines; most of this software currently provides integrative analysis between the data of two or three omics (Table 1.3). Bioinformaticians continue their efforts to provide better algorithms and software, but the variable nature of individual omics data and data properties raise challenges and act as a bottleneck.

TABLE 1.3
Integrative Omics Resources for Agriculture Data Management and Mining

S. No.	Software	Functionality	*Supported Platforms
1	BioCyc/MetaCyc	Metabolic pathway prediction in sequenced genome https://biocyc.org, https://metacyc.org	G|T|P|M
2	C4GEM	Metabolic modeling https://modelseed.org/	T|P|M
3	GIM3E	Flux metabolites prediction https://github.com/brianjamesschmidt/gim3e	T|P
4	IMPaLA	Enrichment analysis; pathway analysis http://impala.molgen.mpg.de	G|T|P|M
5	MapMan4	Enrichment analysis; visualization of data expression https://mapman.gabipd.org	T|P|M
6	Mergeomics	Computational pipeline that integrates multidimensionally, mainly designed for humans. http://mergeomics.research.idre.ucla.edu/	G|T|Eg
7	mixOmics	Data integration; similarity relationship www.mixOmics.org	Mg|T|P|M
8	Omickriging	Data integration and visualization https://github.com/hakyimlab/OmicKriging	T|P|M|F
9	OmicsAnalyzer	Network mapping https://apps.cytoscape.org/apps/omicsanalyzer	T|P|M|F
10	OmicsPLS	Data integration and statistical analysis https://github.com/selbouhaddani/OmicsPLS	Mg|T|P|M
11	OMICtools	Supported Platform for OMICtools	G|T|P|M
12	Paintomics3	Data integration and visualization; pathway analysis and interaction www.paintomics.org	T|P|M
13	PathView	Data integration and visualization https://pathview.uncc.edu/	T|M
14	PlantSEED	Metabolic reconstruction https://modelseed.org/	G|T|M|F
15	Regulomics	Functional characterization of data; retrieve upstream regulators; data mining http://bioinfo.sibs.ac.cn/plant-regulomics/	G|T|

*G: Genomics, T: Transcriptomics, P: Proteomics, M: Metabolomics, Mg; Metagenomics, F: Fluxomics, Eg: Epigenomics

1.6 CHALLENGES IN MULTI-OMICS DATA INTEGRATION

Since the emergence of the concept of multi-omics and systems biology, an enormous amount of data has been and continues to be generated due to the rapid enhancement of technology. Scientists are trying to provide compatible analysis tools and other resources with which to read biological data in a machine-readable interoperable format. Homogenous data from the same experiment can be integrated, whereas heterogenous (KEGG, GO, PPI, microarray etc.) data can be applied based on the meta-analysis (Dupae et al., 2014). For instance, in various environments, how crops modulate their performance can be identified through an integrated 'omics' approach (Fabres et al., 2017). In fact, to some extent it succeeds in this purpose; however, there are a range of challenges that hamper further progress in multi-omics integration and systems biology research (Pinu et al., 2019). The major challenges follow.

1.6.1 Integration of Qualitative and Quantitative Data

Due to the qualitative nature of most of the 'omics' data, it is very hard to reproduce and even harder to compare them. Unavailability of quantitative data forces us to rely on qualitative data only. When only qualitative data are available, multi-omics integration, particularly from multiple sources, becomes difficult (Pinu et al., 2019).

1.6.2 Lack of Standardized Operating Protocols

As most of the omics data is generated through high throughput techniques, different techniques follow different operating protocols. Automatic integration of such data is not a straight forward task and raises the issues.

1.6.3 The Lack of Sufficient Meta-Data

Meta-data provides details about the experiment and is very crucial in data pooling. Lack of sufficient meta-data affects purpose-specific data integration.

1.6.4 Selection of Database and Tools

With the exponential growth of bioinformatics, hundreds of kinds of software and tools are now available for each purpose. However, it is very difficult for a user to differentiate between reliable and non-reliable tools, which can cause confusion in deciding what to select for integration purposes.

1.6.5 Variable Data Formats

Availability of data in different formats, especially in non-standard formats, creates problems in data integration.

1.6.6 Non-Interoperable Tools

Many programs/tools work on non-standard formats and sometimes also provide output data in formats that are incompatible with other programs.

1.6.7 Network Inference and Functional Linkage Network (FLN) Construction

Based on the evidence of various data types, the network topology is constructed (Oliveira, 2019; Wang et al., 2019).

1.7 DATA CHALLENGES IN INTEGRATIVE DATA HANDLING AND MINING

1.7.1 Observation Scale

Per their variable nature, omics data sets share a large variation in the number of observations per sample. Any genome typically includes millions of variants, a transcriptome typically includes tens of thousands of quantified transcripts and proteomes and metabolomes include thousands of quantified molecules (Misra et al., 2018). These wide ranges of scale make it impracticable to normalize and integrate the data for any purpose in a meaningful manner.

1.7.2 Non-Standard Data Identifier

As most of the databases practice their own keywords and IDs for stored data, databases face a lack of standardized data identifier problem. Data mining utilizes these identifiers for data processing and pattern identification as well as face selection/inclusion problems.

1.7.3 Voluminous Data

Integration of multi-omics data results in gigantic data even in a single-purpose integration. If we consider one standard rice NGS experiment, it produces data of a gigabyte (GB) to terabytes (TB) in size. Multi-omics needs to integrate thousands of such data sets from different omics. One can guess how tedious and challenging a task it is to process these data.

1.7.4 Incomplete Data

As agricultural research suffers from a slow speed, comparative tedium, less funds and ignorance, we are far from various bio-molecules and biological processes, especially plant proteins and metabolites. This incomplete picture creates problems and in fact sometimes leads towards false hypotheses and conclusions.

1.7.5 Variety of Data

Like in data integration, many varieties of data also create trouble in data mining. It limits the use of integrated data mining and analytics pipelines.

1.7.6 Dynamic Nature of Data

The dynamic nature of the data and data velocity have a huge impact on data interpretation and decision making. Due to the exponential growth and use of high throughput techniques for omics, it is assumed that in the near future this field will also face the problem of 'concept drift'.

The features discussed in this chapter clearly revealed that mining of integrative data is a very challenging task. The integration of multi-omics data and consideration of a holistic approach towards these data, in terms of understanding systemic interaction between molecules and various biological processes, is not an easy or straightforward task, at least in the near future using currently available data mining facilities. All of these data features jointly raise big data issues, including *volume*, *variety*, *velocity* and *veracity*, and sometimes *value*.

1.8 FUTURE DIRECTIONS FOR INTEGRATION AND MINING OF MULTI-OMICS DATA

1.8.1 Artificial Intelligence

Computational biologists face a big challenge in the near future when it comes to dealing with the gigantic size of integrative data from the practice of multi-omics. Data size and complexity created needs to develop an algorithm and tools to provide expectable support. If we look at the available concepts, AI (artificial intelligence) and big data analytics are both promising and can offer support (Ma et al., 2014); at present, they are already being utilized for different kinds of data problems (e.g., stock exchange, space science, astronomy and marketing). In the last decade, we have witnessed a number of pattern analysis tools based on machine learning (ML) emerging for biological problems, such as ImageJ, PSLIDS, Cell Profiler, Cell Explorer and Cell Tracer (Shamir et al., 2010). In the last few years, computational biologists have also begun to explore the compatibility of ML as well as ANN (artificial neural network) based deep learning for biological data and the possibility of contributing to integrative omics. A few recent studies have demonstrated the potential of deep learning (because of its self-learning capacity, and its handling capacity of unstructured data) in bioinformatics for modeling complex biological data, making new discoveries and decision making (Ma et al., 2014; Oliveira, 2019).

1.8.2 Big Data Analytics

Big data analytics, mainly driven by HPC and specialized software and systems, offer possibilities to deal with integrative omics data and provide a way to process,

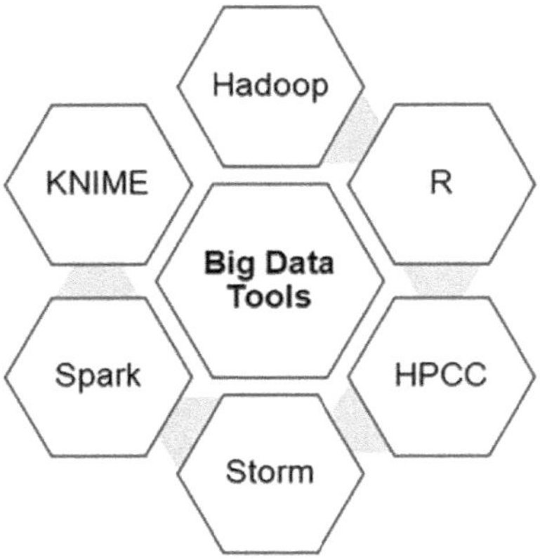

FIGURE 1.3 Open source, big data analytics tools.

mine and interpret data at one go. Big data analytics applications enable users to analyze thriving volumes of data, along with other data that are difficult to handle with conventional analytics methods. A number of general-purpose tools to assist big data analytics are already available (e.g., Hadoop, Yarn, MapReduce, spark, hive etc.; Figure 1.3); they are mainly designed to handle structured as well as unstructured and semi-structured data types. While the use of big data analytics by computational biologists is increasing slowly, especially for agricultural purposes, soon the growth and prospects of integrative omics will result in the development and standardization of bioinformatics-oriented big data analytics tools.

1.8.3 Cloud Computing

Cloud computing provides on-demand availability of computer resources, especially data storage and computing power (Figure 1.4). Cloud computing can provide a solution to 'big data problems' to some extent, as the user can demand any of the available computing facilities, i.e., software-as-a-service (SaaS), infrastructure-as-a-service (IaaS) and platform-as-a-service (PaaS), as required. Use of cloud computing for multi-omics data can provide many advantages to the user as well as the scientific community more broadly. Many of them (1) are less expensive and more cost-effective, (2) make it easy to store huge amount of data, (3) offer easy processing and mining of the data, (4) make it easy to share the data with rest of scientific community, (5) offer easy access the data from anywhere, and (6) enhance collaborative work.

1.9 ADVANTAGE OF INTEGRATIVE OMICS IN AGRICULTURE

Agriculture is very important for humans, as it supplies food and the various raw materials required for daily life. Agriculture is also important for various

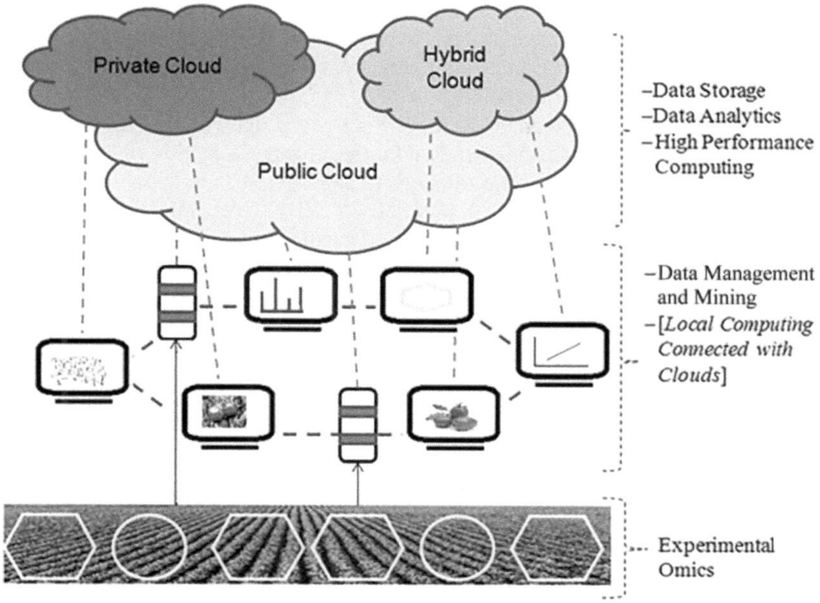

FIGURE 1.4 Cloud computing for multi-omics data (theory).

industries, including grains, vegetables, jute, sugar and a lot more. As central dogma is important and responsible for molecular information flow, agriculture scientists are exploring aspects of all omics to understand the role of various biological molecules in enhancing agriculture products and crop improvement. Most current research has focused on different omics to understand differing insights; as discussed in the chapter, individual omics are not capable of providing a complete picture. Integrative omics provide consideration of multi-omics analysis to understand the molecular/biological process of interest and provide more informative insights that are not accessible through individual omics.

Integrative omics will be helpful in various aspects of agriculture, including:

- Discovery of trait-associated genes and marker development.
- Multi-omics–based breeding design and practice.

REFERENCES

Álvarez-Venegas, R., & De-la-Peña, C. (2016). Editorial: Recent advances of epigenetics in crop biotechnology. *Frontiers in Plant Science*, *7*, 413. https://doi.org/10.3389/fpls.2016.00413

Dixit, S., Srivastava, M., & Katara, P. (2019). Computational assessment of polymorphism with linkage disequilibrium and hotspots of recombination in pathogenic genes of Fusarium oxysporum f. sp. lycopersici. *Legume Research: An International Journal*, *42*(6).

Dupae, J., Bohler, S., Noben, J. P., Carpentier, S., Vangronsveld, J., & Cuypers, A. (2014). Problems inherent to a meta-analysis of proteomics data: A case study on the plants' response to Cd in different cultivation conditions. *Journal of Proteomics*, *108*, 30–54. https://doi.org/10.1016/j.jprot.2014.04.029

Fabres, P. J., Collins, C., Cavagnaro, T. R., & Rodríguez López, C. M. (2017). A concise review on multi-omics data integration for terroir analysis in vitis vinifera. *Frontiers in Plant Science*, *8*, 1065. https://doi.org/10.3389/fpls.2017.01065

Gomez-Casati, D. F., Zanor, M. I., & Busi, M. V. (2013). Metabolomics in plants and humans: Applications in the prevention and diagnosis of diseases. *BioMed Research International*, *2013*, 792527. https://doi.org/10.1155/2013/792527

Jaiswal, P., Cheruku, J. R., Kumar, K., Yadav, S., Singh, A., Kumari, P., Dube, S. C., Upadhyaya, K. C., & Verma, P. K. (2012). Differential transcript accumulation in chickpea during early phases of compatible interaction with a necrotrophic fungus Ascochyta rabiei. *Molecular Biology Reports*, *39*(4), 4635–4646. https://doi.org/10.1007/s11033-011-1255-7

Katara, P. (2014). Potential of bioinformatics as functional genomics tool: An overview. *Network Modeling Analysis in Health Informatics and Bioinformatics*, *3*(1), 1–7.

Ma, C., Zhang, H. H., & Wang, X. (2014). Machine learning for Big Data analytics in plants. *Trends in Plant Science*, *19*(12), 798–808. https://doi.org/10.1016/j.tplants.2014.08.004

Misra, B. B., Langefeld, C. D., Olivier, M., & Cox, L. A. (2018). Integrated omics: Tools, advances, and future approaches. *Journal of Molecular Endocrinology*. https://doi.org/10.1530/jme-18-0055

Oliveira, A. L. (2019). Biotechnology, Big Data and artificial intelligence. *Biotechnology Journal*, *14*(8), e1800613. https://doi.org/10.1002/biot.201800613

Pinu, F. R., Beale, D. J., Paten, A. M., Kouremenos, K., Swarup, S., Schirra, H. J., & Wishart, D. (2019). Systems biology and multi-omics integration: Viewpoints from the metabolomics research community. *Metabolites*, *9*(4). https://doi.org/10.3390/metabo9040076

Rexroad, C., Vallet, J., Matukumalli, L. K., Reecy, J., Bickhart, D., Blackburn, H., Boggess, M., Cheng, H., Clutter, A., Cockett, N., Ernst, C., Fulton, J. E., Liu, J., Lunney, J., Neibergs, H., Purcell, C., Smith, T. P. L., Sonstegard, T., Taylor, J., . . . Wells, K. (2019). Genome to phenome: Improving animal health, production, and well-being – a new USDA blueprint for animal genome research 2018–2027. *Frontiers in Genetics*, *10*, 327. https://doi.org/10.3389/fgene.2019.00327

Shamir, L., Delaney, J. D., Orlov, N., Eckley, D. M., & Goldberg, I. G. (2010). Pattern recognition software and techniques for biological image analysis. *PLoS Computational Biology*, *6*(11), e1000974. https://doi.org/10.1371/journal.pcbi.1000974

Singh, V., Sharma, V., & Katara, P. (2016). Comparative transcriptomics of rice and exploitation of target genes for blast infection. *Agri Gene*, *1*, 143–150.

Van Emon, J. M. (2016). The omics revolution in agricultural research. *Journal of Agricultural and Food Chemistry*, *64*(1), 36–44.

Wang, L., Audenaert, P., & Michoel, T. (2019). High-dimensional Bayesian network inference from systems genetics data using genetic node ordering. *Frontiers in Genetics*, *10*, 1196. https://doi.org/10.3389/fgene.2019.01196

Yadav, A., Vishwakarma, S., Krishna, N., & Katara, P. (2020). Integrative omics: Current status and future directions. In *Recent trends in computational omics: Concepts and methodology*. Nova Science Publishers.

Zhao, C., Zhang, Y., Du, J., Guo, X., Wen, W., Gu, S., Wang, J., & Fan, J. (2019). Crop phenomics: Current status and perspectives. *Frontiers in Plant Science*, *10*, 714. https://doi.org/10.3389/fpls.2019.00714

2 An Overview of Plant–Pathogen Genome Databases and Interaction Resources

S. J. S. Rama Devi, Supriya Babasaheb Aglawe, and Yamini Deepthi

2.1 INTRODUCTION

Plants experience a wide range of biotic stress caused by diverse pathogens like bacteria, viruses, fungi, oomycetes, nematodes etc. These pathogens attack different parts of the plant (root, stem, leaf, flower, fruit) at various stages of the plant life cycle. Plant pathogens can successfully colonise around plant tissues and absorb nutrition from host plants. In response, plants also elicit defence mechanisms against pathogen attacks. Thus, pathogens interfere in plant development and growth and also cause plant death. This results in substantial economic losses in agriculture yield. This is a continuous co-evolutionary arms race between host plants and pathogens and is very well established in agricultural ecosystems, compared to natural ecosystems, due to the constant domestication of agricultural crops (Karasov et al., 2014). Hence, it has become mandatory for the scientific community to work towards acquiring a durable and broad-spectrum host plant resistance. Precise understanding and knowledge of molecular mechanisms and the cellular processes underlying the host resistance genes and virulence effectors of pathogens leading to pathogenicity and their interaction may help to develop disease-resistant plants.

In recent years, various sequencing platforms, such as DNA sequencing (DNA-Seq) [whole genome sequencing (WGS); whole exome sequencing (WES) and epigenome sequencing]; RNA sequencing [whole transcriptome sequencing (WTS); mRNA sequencing (mRNA-Seq)] and small RNA sequencing (Hu et al., 2021), have offered a variety of data to understand the complex functional and comparative genomics of hosts and pathogens. Public data availability enables researchers to elucidate the molecular mechanism of host–pathogen interactions and their co-evolution (Xu & Wang, 2019; Collmer et al., 2009). DNA-Seq is paired with re-sequencing projects of an organism to achieve better molecular insights into the organism and also to aid in comparing diverse genomes. Thus originated the concept of pan-genome (building the species-representative genome to capture distinct sequence variations), and pan-genomes have been initiated for many crop

DOI: 10.1201/9781003153481-2

plants and several pathogen species (Golicz et al., 2016; Chen et al., 2018). Among these projects, some have been completed (Valentin et al., 2021), while others are in the pipeline (https://solgenomics.net/projects/tomato100; www.soybase.org/projects/SoyBase.C2021.01.php), providing enormous datasets for further scientific evaluation (Badet & Croll, 2020). Pan-genome data enables exploration of the genomic diversity of ecotypes and geographical isolates; genome-wide comparative and association studies to identify the genetic elements associated with agronomical important traits (Li & Harkess, 2018); and ecology and evolution of different strains of pathogens (Lücking et al., 2021; Li et al., 2021). RNA-Seq is advantageous for a comprehensive understanding of plant–pathogen interactions, which requires knowledge of associated genes that are differentially expressed in both the host and the pathogen. Dual RNA-Seq technique makes feasible analysing host–pathogen transcriptomes in parallel, which is highly beneficial for monitoring differentially expressed genes accurately, with finer detail assisting in capturing plant-specific transcripts during infection (Westermann et al., 2012; Naidoo et al., 2017). It facilitates the researchers in getting insights into global gene expression changes, including alternative splice variants, and plants' long non-coding RNAs (lncRNAs) under biotic stress (Li, Tai et al., 2018).

This huge collection of sequence data can be analysed using various bioinformatics pipelines. Bioinformatics, an interdisciplinary branch of science, can help us to analyse and interpret a large amount of information generated through sequencing technologies. With recent advancements in sequencing technologies, it is now possible to effectively sequence the entire genome of an organism in a few hours, which has increased the demand for digital databases to collect, store, save and retrieve information according to the need and necessity of the scientific community for further validation and functional characterisation (Zhulin, 2015). In the past decade, various specific and generalised biological databases have been developed and implemented several bioinformatics tools to analyse host–pathogen data in order to understand host–pathogen interactions. The databases facilitate the free flow and faster exchange of information, enabling researchers to find solutions to ever-emerging phytopathogen threats. Open resources maximise the accessibility of the information towards future investigation.

The current chapter aims to highlight how genomic technologies are exploited for understanding various plant–pathogen interactions. The information compiled here could also serve as a valuable resource for plant pathologists and molecular biologists.

2.2 PLANT GENOME DATABASES – AN OVERVIEW

Recent advancements in the field of genomics, such as high-density genotyping and next-generation technologies (NGS), have made it possible to sequence the whole genomes of several plants of economic importance, such as cereal crops like rice (Eckardt, 2000); barley (Liu et al., 2020); wheat (Ling et al., 2013; Bierman, 2017); vegetables like radish (Hirakawa, 2017); commercial crops like cotton (Li, Fan et al., 2014a); and sugarcane (Riaño-Pachón & Mattiello, 2017), as well as ornamental plants (Yagi, 2018). To date, more than 300 plant species genomes have been sequenced successfully and are available publicly for future exploration and data-driven research (Shirasawa et al., 2021). As of writing, the NCBI, which is a major curator for several genomes, has 254 reference sequence category representative genomes of land plants (www.ncbi.nlm.nih.gov/genome/browse#!/eukaryotes/). Similarly, TAIR (www.arabidopsis.org/); RAPDB (https://rapdb.dna.affrc.go.jp/); and IWGSC (www.wheatgenome.org/) maintain the reference genome sequences of and several analysis tools for the *Arabidopsis thaliana*, *Oryza sativa* and *Triticum aestivum* genomes, respectively. Comparative genomics studies relationships among species for their evolutionary and functional genomics, and databases with integrated bioinformatics tools have facilitated comparative genomics study, such as the Gramene database (www.gramene.org/), which provides the comparative sequence information of several grass genomes. The genome sequencing, genome re-sequencing, pan-genomics and transcriptomics projects would eventually aid in genome variation analysis among species/cultivars; lead to novel resistance gene discovery in host plants; and aid in functional analysis of virulence genes leading to pathogenicity. These efforts will accelerate the crop improvement process for combating disease resistance.

Plant immunity is triggered by using the host's pattern recognition receptors (PRRs) to recognise the molecular patterns associated with pathogens or microorganisms (PAMPs and MAMPs). If pathogens exploit the effector proteins for a compatible reaction, the resistance (R) genes derived from R proteins will identify them as harbouring vertical resistance to plants (Silva & Micheli, 2020). The following section discusses the important online resources that aid in generating new biological insights for resistance crop breeding.

The Plant Resistance Genes database (PRGdb 3.0): This is an online resource that maintains 153 plant resistance genes (*PRGs*) and 23 avirulence (*Avr*) genes from 90 species. The user interface DRAGO2 pipeline can predict putative resistance genes from the transcriptome or proteome data. The user can access the information available using the drop-down menu, which includes reference and putative *PRGs*, *Avr* genes and all resistance genes. This web resource is highly useful to the plant science research community and plant breeders working towards disease resistance (http://prgdb.org/prgdb/; Osuna-Cruz et al., 2018).

RRGPredictor: This is a more efficient and effective tool to predict plants' PRRs and R proteins domains by using tab-separated value (tsv) files without sequence homology. The prediction is based on the protein domain arrangements within the sequence. This tool is very sensitive, with low computational cost when compared to other R gene predictive tools like DRAGO2 and RGAugury (https://github.com/RanerJSS/RRG_predictor; Silva & Micheli, 2020).

Plant stress RNA-seq Nexus (PSRN): This comprehensive database (includes 26 plant stress stress-derived RNA-Seq datasets, 12 plant species and 97 samples) includes stress-specific differentially expressed genes. It provides expression datasets of coding and lncRNAs (long non-coding RNA) and identifies differentially expressed transcripts between different stress subset datasets (http://syslab5.nchu.edu.tw/PSRN; Li, Liu et al., 2018).

In addition to the above, Table 2.1 summarises the list of important species-specific, integrated and pan-genome plant genome databases and also

TABLE 2.1
List of Various Online Plant-Specific Resources

Sl. No.	Database	Year	URL	Description	Reference
1	Tomato genome (*Solanum lycopersicum*) by Solanaceae Genomics Network	2012	https://solgenomics.net/organism/Solanum_lycopersicum/genome	Reference tomato genome sequence	The Tomato Genome Consortium, 2012
2	Potato genome (*Solanum tuberosum*) by Solanaceae Genomics Network	2011	https://solgenomics.net/organism/Solanum_tuberosum/genome	Reference potato genome sequence	Potato Genome Sequencing Consortium, 2011
3	Hot pepper genome (*Capsicum annuum*) by Solanaceae Genomics Network	2014	https://solgenomics.net/organism/Capsicum_annuum/genome	Reference pepper genome sequence	Kim et al., 2014
4	Common tobacco (*Nicotiana tabacum*) by Solanaceae Genomics Network	2014	https://solgenomics.net/organism/Nicotiana_tabacum/genome	Reference tobacco genome sequence	Sierro et al., 2014
5	Cotton genome (*Gossypium hirsutum*) by CottonGen database	2014	www.cottongen.org/tools/jbrowse	Genomics, genetics and breeding database of cotton	Li, Fan et al., 2014
6	Sugarcane genome (*Saccharum officinarum*) by Sugarcane Genome Hub	2018	https://sugarcane-genome.cirad.fr/	Sugarcane genomic database	Garsmeur et al., 2018
7	Maize genome (*Zea mays*) database	2019	www.maizegdb.org/	Maize multi-genome genetics and genomics database	Portwood et al., 2019
8	Brinjal (*Solanum melongena L*) – eggplant genome database	2014	http://eggplant.kazusa.or.jp/	Draft genome dataset for egg plant brinjal	Hirakawa et al., 2014
9	Soybean (*Glycine max*) – SoyBase	2021	https://soybase.org/search/	Integrated soybean genomic and genetic database; visualisation of soybean pan-genome	Brown et al., 2021

Integrated Databases for Comparative Plant Genomics					
1	Plant-ImputeDB	2021	http://gong_lab.hzau.edu.cn/Plant_imputeDB/#!/	Genotype imputation using high-quality reference sequence panels	Y. Gao et al., 2021
2	Plant Genome Database Japan (PGDBj)	2013	http://pgdbj.jp/?ln=en	Integrated of plant genome-related databases	Nakaya et al., 2017
3	Phytozome	2012	www.plantgdb.org/	Comparative plant genomics portal	Goodstein et al., 2012
4	PGSB PlantsDB	2016	http://pgsb.helmholtz-muenchen.de/plant/index.jsp	Integrated platform for analysing the various plant genomes using bioinformatics tools	Spannagl et al., 2016
Plant Pan-genome Projects					
1	Wheat (*Triticum aestivum*) 10+ Wheat Genomes Visualizer	2020	https://kiranbandi.github.io/10wheatgenomes/	Comparative synteny analysis in wheat	Walkowiak et al., 2020
2	Wheat (*Triticum aestivum*) 10+ Wheat Genomes Project	2020	www.crop-haplotypes.com/Wheat/haplotype/1A	Comparative haplotype analysis in wheat	Walkowiak et al., 2020
3	Rice (*Oryza sativa*) Rice SNP-Seek Database	2017	https://snp-seek.irri.org/index.zul;jsessionid=EC95D355666E5E3C84554DAAB6DFB7AA	Genotype, phenotype, and variety information for 3010 rice accessions	Li, Wang et al., 2014b
4	Rice (*Oryza sativa*) RPAN	2016	https://cgm.sjtu.edu.cn/3kricedb/ (or) www.rmbreeding.cn/pan3k	Rice pan-genome browser	Du et al., 2017
5	GreenPhylDB v5		www.greenphyl.org/cgi-bin/index.cgi	Contains 19 pan-genomes of five crops (rice; maize; banana; grape and cocoa)	Valentin et al., 2021
6	Rice Galaxy	2019	https://github.com/InternationalRiceResearchInstitute/RiceGalaxy	Publicly available big datasets of rice	Juanillas et al., 2019
7	The SOL-100 sequencing project	Ongoing	https://solgenomics.net/organism/sol100/view	Creation of a common genomic framework that includes sequences and phenotypes of 100 Solanaceae genomes	

(Continued)

TABLE 2.1 *(Continued)*
List of Various Online Plant-Specific Resources

Sl. No.	Database	Year	URL	Description	Reference
8	Tomato (*Solanum lycopersicum*) 100 tomato genome project	Ongoing	https://solgenomics.net/projects/tomato100	Aims to generate genome sequence from 100 diverse tomato accessions	
9	PanSoy	Ongoing	www.soybase.org/projects/SoyBase.C2021.01.php	Pan-genome of 204 elite cultivars of soybean	
Plant Disease Resistance Study Utility Bioinformatics Tools					
1	Meta-i6mA	2021	http://kurata14.bio.kyutech.ac.jp/Meta-i6mA/	Integrated platform for identifying the interspecies DNA N6-methyladenine sites of plant genomes	Hasan et al., 2021
2	NLGenomeSweeper	2020	www.mdpi.com/2073-4425/11/3/333/s1	Automated accurate identification of plant NBS-LRR candidate genes without previous gene predictions	Toda et al., 2020
3	RGAugury	2016	https://bitbucket.org/yaanlpc/rgaugury.	Online bioinformatics tool for genome wide identification of resistance gene analogues (RGA)	Li et al., 2016
4	RefPlantNLR	2020	https://itol.embl.de/shared/JKourelis	Comprehensive reference dataset for experimentally validated plant nucleotide-binding leucine-rich repeat (NLR) immune receptors	Kourelis & Kamoun, 2020
5	NLR-Annotator	2020	https://github.com/steuernb/NLR-Annotator	A de novo annotation tool for plant nucleotide-binding leucine-rich repeat (NLR) genes	Steuernagel et al., 2020
6	DRPPP (Disease resistance plant protein predictor)	2016	http://14.139.240.55/NGS/download.php	Tool for the prediction of disease resistance proteins in plants	Pal et al., 2016
7	prPred	2021	https://github.com/Wangys-prog/prPred	Prediction tool to identify the plant resistance proteins	Wang et al., 2021

8	NBSPred	2016	http://soilecology.biol.lu.se/nbs/	This pipeline differentiates the NBS-LRR genes from other non–NBS-LRR genes utilising genome, transcriptome and proteome datasets	Kushwaha et al., 2016
Plant–Pathogen Interactions					
1	InterSPPI	2019	http://systbio.cau.edu.cn/intersppi/index.php	Comprehensive and experimentally validated prediction tool for plant pathogen protein–protein interactions	Yang et al., 2019
2	PredHPI	2021	http://bioinfo.usu.edu/PredHPI/	Integrated framework for the detection and visualisation of plant pathogen network interactions	Loaiza et al., 2021
3	HPIDB 2.0	2016	www.agbase.msstate.edu/hpi/main.html	Resource for host–pathogen interactions with 45238 manually curated entries	Ammari et al., 2016
4	VirusMentha	2015	http://virusmentha.uniroma2.it/	Resource for plant virus protein–protein interactions	Calderone et al., 2015
5	PHYTOPATH	2016	www.phytopathdb.org	Integrated resource for plant–pathogen interactions	Pedro et al., 2016
6	ApoplastP	2017	http://apoplastp.csiro.au/	Predicts apoplastic localisation of plant apoplastic proteins as well as pathogen-derived apoplastic effectors	Sperschneider et al., 2018b
7	NEMATIC (NEMatode–Arabidopsis Transcriptomics Interaction Compendium)	2014	www.uclm.es/grupo/gbbmp/english/nematic.asp	Pipeline for the analysis of plant–nematode interactions	Cabrera et al., 2014

provides a summary of relevant online resources available for improving plant disease resistance.

2.3 PHYTOPATHOGENIC BACTERIA DATABASES

Phytopathogenic bacteria are diverse and have a close relationship with plants. The first plant pathogenic bacteria to be sequenced was *Xylella fastidiosa*, which has a wide host range and is considered one of the major destructive bacteria across the globe (Simpson et al., 2000; Schneider et al., 2020). Subsequently, with the exponential increase in sequencing technologies, many phytopathogenic bacterial genome sequences have been sequenced successfully and published (Collmer et al., 2009). There are currently several microbial databases based on model organisms, genomic resources, metagenomics, protein families, etc., which are well summarised (Zhulin, 2015). Furthermore, the database mining of genomes of plant pathogenic bacteria has also been excellently reviewed (Vinatzer & Yan, 2008). The functionality of the foremost microbial genome web portals was well compiled for their utility functionality and user friendliness; however, these databases are not exclusive to plant pathogens. These bacterial genomic or any other web resources are very useful for carrying gene annotations; performing comparative synteny analysis; browsing genes and protein functional analysis; deciphering promoter architecture, etc. (Karp et al., 2019). Straub et al. (2021) updated the population genomics of phytopathogenic bacteria to better understand their origin, transmission during outbreaks and the genetic basis of host range, virulence and evolution. In addition, the pan-genomic type III effector database was developed from 12 phytopathogenic Ralstonia strains known to cause vascular bacterial wilt with a wide host range (Sabbagh et al., 2019). Recent phytopathogenic bacteria databases have been tabulated in Table 2.2.

2.4 FUNGI DATABASES

Fungal pathogens pose the greatest threat to world crop production, leading to significant yield losses annually. The first fungal genome of *Saccharomyces cerevisiae* (yeast) was sequenced in 1996 (Goffeau et al., 1996). Since then, genome sequencing has enabled researchers to understand the remarkable variability of genome size and architecture of fungi. By exploiting the genomic information, it is known that fungal virulence or effector genes are not randomly localised but are rather associated closely with repetitive and transposable elements (Dong et al., 2015a). Möller and Stukenbrock (2017) critically reviewed the genome architecture and evolution of plant fungal pathogens. Furthermore, high throughput DNA sequencing technologies have significantly transformed fungal diversity studies (Baldrian et al., 2021) and contributed to pan-genome sequences of many plant pathogenic fungi (Xu et al., 2006; Badet & Croll, 2020). It is predicted that the amalgamation of big data from various independent research may open new horizons in future mycological studies (Baldrian et al., 2021). The 1K fungal genome

TABLE 2.2
Bioinformatics Analysis Tools Available for Varied Phytopathogens

Sl. No.	Database	Year	URL	Description	Reference
Phytopathogenic Bacteria Bioinformatics Resources					
1	GOLD – Genomes Online Database	2021	https://gold.jgi.doe.gov/	Frequently updated comprehensive database for genome and metagenomics projects worldwide (which includes bacteria and viruses)	Mukherjee et al., 2021
2	Bacterial Effector Analyzer 2.0 (BEAN 2.0)	2015	http://systbio.cau.edu.cn/bean/	Database, prediction and analysis for type III secreted effectors of Gram-negative pathogenic bacteria	Dong et al., 2015b
3	PGAweb	2018	http://PGAweb.vlcc.cn/	Bacterial pan-genomic analysis	Chen et al., 2018
4	VRprofile	2018	http://bioinfo-mml.sjtu.edu.cn/VRprofile	Web server used to identify and localise virulence as well as antibiotic resistance genes from pathogenic bacteria genomes	Li, Tai et al., 2018
5	gDAT (graphical downstream analysis tool)		https://github.com/ut-planteco/gDAT	Analysis of rDNA sequences	Vasar et al., 2021
6	The Pathosystems Resource Integration Center (PATRIC)	2021	https://patricbrc.org/	Integrated bacterial bioinformatics and analysis tool for varied datasets like genomics; metagenomics; transcriptomics; protein–protein interactions	Wattam et al., 2014; Parrello et al., 2021
7	Microbial Genome Database (MBGD)	2018	http://mbgd.genome.ad.jp/	Comprehensive analysis of orthology relationships from bacteria and other microorganisms	Uchiyama et al., 2018
8	Virulence Factor Database (VFDB)	2016	www.mgc.ac.cn/VFs/	Comprehensive repository for bacterial virulence factors attacking plants and animals	Liu et al., 2019

(Continued)

TABLE 2.2 *(Continued)*
Bioinformatics Analysis Tools Available for Varied Phytopathogens

Sl. No.	Database	Year	URL	Description	Reference
Plant Parasitic Fungal Databases					
1	GlobalFungi	2021	https://globalfungi.com	Web interface to access and visualise the high throughput fungal sequencing data from various geographical regions, ecosystems and habitats	Větrovský et al., 2020
2	FungiDB	2021	https://fungidb.org/fungidb/app	Integrated functional genomics platform for fungi and oomycete	Basenko et al., 2018
3	FUNYBASE (FUNgal phYlogenomic dataBASE)	2008	http://genome.jouy.inra.fr/funybase	Performs comparative and phylogenomic study in fungi	Marthey et al., 2008
4	Predector	2021	https://github.com/ccdmb/predectordata	Prediction of plant fungal effector proteins	Jones et al., 2021
5	EffectorP 2.0	2018	http://effectorp.csiro.au	Prediction of high confidence effector proteins secreted by plant pathogenic fungi	Sperschneider et al., 2018a
6	Funeffector-Pred	2020	http://lab.malab.cn/~wangchao/softwares/software.html	Prediction of fungal effector proteins	Wang et al., 2020
7	Localiser	2017	http://localizer.csiro.au/	Prediction of sub-cellular localisation (mitochondria; chloroplast and nuclei) of plant- and pathogen-derived effectors	Sperschneider et al., 2017
8	FungSatDb	2021	http://webtom.cabgrid.res.in/fungsatdb/	Largest fungal microsatellite database based on whole genome sequencing data of >1900 fungal species/strains	Iquebal et al., 2021
9	MycoCosm	2014	https://mycocosm.jgi.doe.gov/mycocosm/home	Web interface for genomic data integration and analysis of fungi; also facilitates the community participation through nomination of new fungal species	Grigoriev et al., 2014

Plant Parasitic Nematode Related Databases					
1	PPNID	2020	https://github.com/xueqing4083/PPNID	Reference database and automated software pipeline for rapid identification of plant parasitic nematodes	Qing et al., 2020
2	Global database for soil nematodes	2020	https://github.com/hooge104/2020_global_nematode_dataset	Comprehensive dataset for global soil nematode abundance that includes 6825 samples	van den Hoogen et al., 2020
3	WormBase ParaSite	2017	http://parasite.wormbase.org	A comprehensive online resource for helminthes genomics, which also includes PPN	Howe et al., 2017; Bolt et al., 2018
4	NemChR-DB	2021	http://ohalloranlab.net/nemchr-db.	Database for parasitic nematode chemoreceptors	Langeland et al., 2021
5	SCNBase (soybean cyst nematode)	2019	SCNBase.org	Integrated genomic and transcriptomics data for soybean cyst nematode *Heterodera glycines*	Masonbrink et al., 2019
6	Nematode.net v 4.0	2014	http://nematode.net	Includes plant parasitic genomics information	Martin et al., 2015
Plant Parasitic Oomycetes Related Databases					
1	Oomycete Gene Order Browser (OGOB)	2019	https://ogob.ie	Database for comparative genomic and syntenic analyses of oomycete genomes	McGowan et al., 2019
2	Phytophthora-ID	2011	http://phytophthora-id.org	Dedicated web resource for *Phytophthora* species and genotypes identification	Grünwald et al., 2011
3	OomyceteDB	Yet to publish	http://oomycetedb.cgrb.oregonstate.edu	Provides oomycetes species identification based on Rps10 (Ribosomal Protein S10) barcoding	http://oomycetedb.cgrb.oregonstate.edu/about.html
4	Oomycete transcriptomics database (OTD 4.0)	2012	www.eumicrobedb.org/transcripts/	Integrated resource for oomycetes transcriptomics data	Tripathy et al., 2012

(Continued)

TABLE 2.2 *(Continued)*
Bioinformatics Analysis Tools Available for Varied Phytopathogens

Sl. No.	Database	Year	URL	Description	Reference
5	Oomycete Gene Table	2019	www.sbi.kmutt.ac.th/cgi-bin/gt/viewer?organism=oomycetes&build= 190208	Online resource for oomycetes comparative genomic analysis	Rujirawat et al., 2019
6	EffectorO	2021	https://bremia.ucdavis.edu/effectorO.php	In silico analysis pipeline for the prediction of oomycetes secreted effectors using genomes	Nur et al., 2021
Plant Virus Related Databases					
1	VirAmp	2015	https://galaxyproject.org/use/viramp/	Virus genome assembly, analysis with high coverage of NGS data	Wan et al., 2015
2	Virus–host DB	2016	www.genome.jp/virushostdb/majorhost/plant	Organised database for host and virus relationship	Mihara et al., 2016
3	Descriptions of Plant Viruses (DPV)	2006	www.dpvweb.net/	Online resource for plant viruses	Adams & Antoniw et al., 2006
4	Indian Plant Virus database	2017	http://14.139.189.27/virusdb/	Assembled information of plant viruses limited to India	Bhat et al., 2017
5	VirusDetect	2017	http://virusdetect.feilab.net/cgi-bin/virusdetect/index.cgi	Virus identification using small RNA dataset	Zheng et al., 2017

sequencing project may provide new clues for fungal diversity and associated pathogenesis (http://1000.fungalgenomes.org/home/about/; https://mycocosm.jgi.doe.gov/mycocosm/home/1000-fungal-genomes). To date, there are several web resources available to conduct comparative functional genomic data analysis and visualisation. Table 2.2 summarises the most recent and relevant fungal databases, which are important for understanding fungal pathogenesis.

2.5 PLANT PARASITIC NEMATODE DATABASES

Plant parasitic nematodes (PPNs) comprising more than 4100 species are considered a major threat to sustainable agriculture, leading to significant economic losses annually (Phani et al., 2021). Most PPNs are obligate biotrophs that attack and feed on the host plant root system by either vascular tissue pericycle (root-knot and cyst nematodes) or direct feeding on cortical cells, leading to stunted growth. Jones et al. (2013) surveyed and ranked the top ten plant parasitic nematodes based on scientific and economic importance. The nematode genome sizes vary and have different ploidy levels, making it difficult to carry on comparative genomics. The first PPN sequenced genomes were *M. incognita* (Abad et al., 2008) and *M. hapla* (Opperman et al., 2008), which revealed the peculiar genome architecture of PPNs (Jones et al., 2013). Recently, the genome of *M. graminicola* was sequenced employing whole genome sequencing, which may serve as a molecular resource to perform comparative genomics of *Meloidogyne* species and its close relatives (Phan et al., 2020). The genomics approach was utilised, and transposable elements were identified as being responsible for the genome plasticity of root-knot nematode *Meloidogyne incognita* (Kozlowski et al., 2020). More than ten PPN genomes were sequenced and submitted to WormBase ParaSite (Montarry et al., 2021), and recently, the nematode chemoreceptor database was developed by exploiting the available genome sequences (Langeland et al., 2021). Kikuchi et al. (2017) summarised the genome evolution of PPNs, where the author discussed the genome statistics, polyploidy, genomics underlying effector genes and reproduction parthenocarpy, and also compared differentially expressed genes during plant nematode interactions. Likewise, Ali et al. (2015) compiled the plant nematode interactions and summarised that PPNs encode majorly for cell wall degrading enzymes and also discussed the secretome (secreted proteins) and effector genes that are key players for parasitism. In addition, the various biotechnological approaches employed for understanding the nematode effector proteins, protein–protein interactions, protein–DNA interactions, expression analysis, and the role of pheromone conversation during plant nematode interactions was critically reviewed (Eves-van den Akker, 2021; Eves-van den Akker et al., 2021). Omics technologies, it is suggested, can be deployed to precisely understand the plant and root-knot nematode interactions (Ibrahim et al., 2019). However, nematode-specific omics databases are underrepresented, which can be attributed to limited genomic resources that need to be focused on in the near future (Powers et al., 2021). Table 2.2 details the important online resources available for PPNs.

2.6 OOMYCETES DATABASES

Oomycetes, fungal-like filamentous eukaryotic microorganisms, are the most devastating plant pathogens, such species as *Phytophthora infestans* (late blight of potato), *Pythium* (causes damping off in seedlings) representing the biggest threat to global food security. Oomycetes are considered an emerging threat to agriculture with a significant impact (Derevnina et al., 2016). The success of oomycetes lies in their broad host range and capacity to overcome host resistance. With the help of genomics, researchers could identify the repertoires of genes that encode virulence effectors. The advent of genomics tools using sequencing technologies pioneered the understanding of oomycetes' pathogenicity and infection mechanisms. The comparative genome analysis provided new insights into the virulence genes, pathogenesis evolution and phytopathogenic oomycetes' survival mechanisms (Adhikari et al., 2013). Kamoun et al. (2015) surveyed the international scientific literature and listed the top ten destructive phytopathogen oomycetes. It is believed that genomic databases are required as the most effective strategies to tackle the emergence of oomycete threats (Derevnina et al., 2016). McCarthy and Fitzpatrick (2017) conducted the first large-scale whole genome phylogenomic analysis of 37 oomycetes genomes. McGowan et al. (2020) reviewed the recent advances in oomycetes genomics and discussed the insights gained by merging genomic, transcriptomic and proteomic analysis towards oomycetes genome organisation and infection mechanisms. Also, the authors updated the phylogeny of 65 oomycetes genomes using a phylogenomic approach. Comparative genomics of *Phytophthora* genomes identified the genomic features associated with pathogenicity, and horizontal gene transfer event from red algae (R. F. Gao et al., 2021). Table 2.2 summarises the various online databases available for understanding the oomycetes genome architecture, transcriptome and effector protein analysis.

2.7 VIRUS DATABASES

Plant viruses cause serious threats and considerable economic losses for sustainable agriculture. This can be attributed to the genome diversity of phytoviruses and their capacity to evolve rapidly and dynamically (Farooq et al., 2021). Phytoviruses can cause enormous modifications, like metabolic and physiological changes, within the host plant; in this scenario, the plant cell wall plays a key role in deciding the plant resistance or susceptibility (Kozieł et al., 2021). Madronero et al. (2019) summarised the characteristics of viral genomes of Colombia. The authors also discussed the proteomics application, bioinformatics tools and omics databases for plant virus interaction studies. Souza et al. (2019) highlighted the applications of proteomic approaches in understanding plant–virus interactions and also emphasised the changes accumulated in photosynthesis-related proteins during plant–virus interactions. High-throughput sequencing (HTS) technologies coupled with omics technologies constitute a powerful toolbox for novel virus discovery and deciphering plant viral interactions, and several analysis pipelines are available for virus detection using HTS data (Kutnjak et al., 2021). Developed knowledge from publicly available resources can be used to create new diagnostic methods for emerging diseases. A recent review has very efficiently discussed the magnitude of viral metagenomics in addressing the origin of plant viruses and the interaction of viruses with host and transmission vectors (Lefeuvre et al., 2019). A selected list of phytoviruses databases is tabulated in Table 2.2.

2.8 PLANT–PATHOGEN INTERACTION GENOME DATABASES

Plant–pathogen interactions are host-specific, complex and fine-tuned molecular communications between the host and pathogen. The molecular perspective of plant–pathogen interactions was well reviewed recently (Kumar & Nanda, 2021). Similarly, Li and Zhang (2016) summarised the plant–pathogen interactions from protein–protein interaction networks perspective and tabulated the available bioinformatics tools. The authors opined that it is a need of the hour to further investigate the Big Data to understand plant–pathogen interactions from a systems biology perspective and to identify potential candidate genes. Imam et al. (2016) reviewed the host–pathogen interactions and tabulated the established plant–pathogen systems. It is to be noted that, in the post-genomics era, only eight well-established model pathosystems are available to understand the plant–pathogen interactions. The research gap in this direction has to be focused by the research community to establish the diverse pathosystems with the help of omics tools, for which the available genomics information would be highly beneficial (Figure 2.1). This is in agreement also with the opinion of Peyraud et al. (2017), who critically discussed and reviewed molecular plant–pathogen interactions from a systems biology perspective. In this context, the following section outlines some available databases developed for plant–pathogen interaction.

PathoPlant: (www.pathoplant.de; Bülow et al., 2007) This is an online web resource meant to analyse plant–pathogen interactions and the signal transduction pathways involved in plant pathogenesis. This database contains information about 99 plants and 107 plant pathogens (including bacteria, fungi, viruses and nematodes) available to study 7692 genes, 350 interaction networks, 173 stimuli, 619 references and 2588 hyperlinks. In addition, *in silico* expression analysis can be performed. This database is linked with publicly available popular big databases such as GenBank, PubMed, SWISS-PROT etc. However, it has not been updated after 2016.

Pathogen Host Interaction database (PHI-base): (www.phi-base.org/releaseNote.htm) PHI-base is an online interactive host–pathogen database integrated with genomics information for several experimentally verified host and pathogen genes to interpret interaction networks. PHI-base maintains the information of experimentally verified pathogen effector genes, virulence and pathogenicity genes from fungi, oomycete and bacterial pathogens that infect plants, animals and insect hosts. This active database has been updated from time to time since 2005. The latest release, version 4.10; 2020, has updated with 15928 interactions between host and pathogens (Urban et al., 2020).

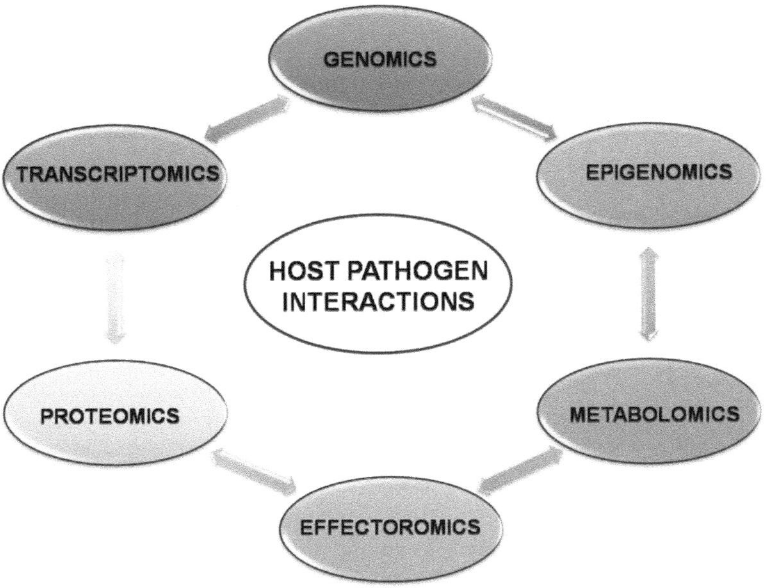

FIGURE 2.1 The illustration depicting the most effective omics technologies-based strategy for better understanding of host–pathogen interactions, where bioinformatics tools play a fundamental role for their integration.

Hemiptera Phytoplasma Plant (HPP) database: (http://trivellone.speciesfile.org/index.asp) This is an online web portal that facilitates biological interactions between the vector (Hemiptera) – pathogen (phytoplasma) and the host plant. Phytoplasmas (without cell wall bacteria, which are obligate parasites on plant phloem sap) are responsible for varied plant diseases across different crops worldwide (Trivellone, 2019).

OmicsDB: Pathogens: (http://pathogens.omicsdb.org; Hansen & Olsson, 2020) This is a publicly available database that gives access to genomics, transcriptomics datasets and gene co-expression networks for selected pathogens and reference species and also provides facilities for various analysis tools (viz. BLAST, Cytoscape etc.). It also has co-expression networks for ten species, comprising four pathogens and six references. User can browse data and analyse and compare interaction networks across the species. The output information offers visualisation graphs, which are easy to understand and flexible for querying.

PlaD: (http://systbio.cau.edu.cn/plad/index.php or http://zzdlab.com/plad/index.php) This gives access to enormous, publicly available gene expression data and information about transcriptional reprogramming, which is a central molecular process during plant–pathogen interactions. This database contains curated gene expression of 2444 pathogenesis-related genes from four model plants viz., *Arabidopsis*, rice, maize and wheat and also offers user-friendly visualisation of expression profile. This web resource provides an excellent platform to explore co-regulated genes during infection initiation and progression for various pathogens (Qi et al., 2018).

2.9 FUTURE PERSPECTIVES AND CONCLUDING REMARKS

Revolutionary advancements in sequencing technologies led to the development of abundant bioinformatics databases to effectively understand the complex biology and evolution of hosts and pathogens. Bioinformatics analysis tools assist the research community in framing the hypothesis before the wet lab experimentation, which is a cost-effective and time-saving strategy. Currently, several databases are available that are designed for genomics study along with other integrated analysis tools for host and pathogen. The present chapter aimed to explore the available comprehensive web resources of plants, phytopathogens and their interactions. The following section emphasises some existing knowledge gaps that need to be addressed to widen the scope, application and efficiency to better understand the ever-emerging warfare between host plants and pathogens.

- In a holistic system biology approach, integration of omics data (genomics, transcriptomics, proteomics, metabolomics, phenomics, secretomics, effectoromics) is a powerful strategy to delineate the complexity of phytopathogen interactions and virulence mechanisms for effective disease control and prevention (Figure 2.1; Arroyo-Velez et al., 2020; Castro-Moretti et al., 2020; Xu & Wang, 2019).
- Spatio-temporal, tissue-specific and parallel evaluation of co-expression studies between host and pathogen during infection and disease progression are valuable resources for researchers to study the multifaceted dynamics of host–pathogen interactions. However, some scientific reports stated that these studies must be extended beyond model crops and pathogens.
- Regular updating of databases is essential. It is advisable to design the automation of algorithms to acquire the data and literature from public databases (such as NCBI, Swiss-Prot, PDB, Ensemble etc.). This would be a cost-effective strategy to maintain the database with minimum effort and human interference for their long-term utility.
- To increase the scope, popularity and widespread application of databases, the developer should prefer an easy, user-friendly interface and visualisation tools since most molecular biologists have only basic knowledge of command-based computer languages.
- Recent scientific evidence has revealed that abiotic stress may also trigger plant susceptibility towards biotic stresses (Zarattini et al., 2021; Desaint et al., 2021). In this perspective, efforts should be directed towards creating a database that integrates knowledge of abiotic stresses, climate change and its role or influence over host–pathogen biology, evolution and interactions. No such databases are available at present. This research gap must be filled in the near future since it is inevitable that climate-resilient crops will be developed to feed an ever-increasing global population with limited natural resources and varied climate changes.

REFERENCES

Abad, P., Gouzy, J., Aury, J. M., Castagnone-Sereno, P., Danchin, E. G., Deleury, E., Perfus-Barbeoch, L., Anthouard, V., Artiguenave, F., Blok, V. C., Caillaud, M. C., Coutinho, P. M., Dasilva, C., De Luca, F., Deau, F., Esquibet, M., Flutre, T., Goldstone, J. V., Hamamouch, N., . . . Wincker, P. (2008). Genome sequence of the metazoan plant-parasitic nematode Meloidogyne incognita. *Nature Biotechnology*, *26*(8), 909–915.

Adams, M. J., & Antoniw, J. F. (2006). DPVweb: A comprehensive database of plant and fungal virus genes and genomes. *Nucleic Acids Research*, *34*(suppl_1), D382–D385.

Adhikari, B. N., Hamilton, J. P., Zerillo, M. M., Tisserat, N., Lévesque, C. A., & Buell, C. R. (2013). Comparative genomics reveals insight into virulence strategies of plant pathogenic oomycetes. *PloS One*, *8*(10), e75072.

Ali, M. A., Abbas, A., Azeem, F., Javed, N., & Bohlmann, H. (2015). Plant-nematode interactions: From genomics to metabolomics. *International Journal of Agriculture and Biology*, *17*(6).

Ammari, M. G., Gresham, C. R., McCarthy, F. M., & Nanduri, B. (2016). HPIDB 2.0: A curated database for host – pathogen interactions. *Database*. https://doi.org/10.1093/database/baw103

Arroyo-Velez, N., González-Fuente, M., Peeters, N., Lauber, E., & Noël, L. D. (2020). From effectors to effectomes: Are functional studies of individual effectors enough to decipher plant pathogen infectious strategies? *PLoS Pathogens*, *16*(12), e1009059.

Badet, T., & Croll, D. (2020). The rise and fall of genes: Origins and functions of plant pathogen pangenomes. *Current Opinion in Plant Biology*, *56*, 65–73.

Baldrian, P., Větrovský, T., Lepinay, C., & Kohout, P. (2021). High-throughput sequencing view on the magnitude of global fungal diversity. *Fungal Diversity*, 1–9.

Basenko, E. Y., Pulman, J. A., Shanmugasundram, A., Harb, O. S., Crouch, K., Starns, D., Warrenfeltz, S., Aurrecoechea, C., Stoeckert, C. J., Kissinger, J. C., Roos, D. S., & Hertz-Fowler, C. (2018). FungiDB: An integrated bioinformatic resource for fungi and oomycetes. *Journal of Fungi*, *4*(1), 39.

Bhat, A. I., Martin, A., John, A., Isaac, N., & Eapen, S. J. (2017). Indian plant virus database – a platform for showcasing research on plant viruses in India. *Current Science*, *113*(1), 27–29.

Bierman, A., & Botha, A. M. (2017). A review of genome sequencing in the largest cereal genome, Triticum aestivum L. *Agricultural Sciences*, *8*(2), 194.

Bolt, B. J., Rodgers, F. H., Shafie, M., Kersey, P. J., Berriman, M., & Howe, K. L. (2018). Using WormBase ParaSite: An integrated platform for exploring helminth genomic data. In *Eukaryotic genomic databases* (pp. 471–491). Humana Press.

Brown, A. V., Conners, S. I., Huang, W., Wilkey, A. P., Grant, D., Weeks, N. T., Cannon, S. B., Graham, M. A., & Nelson, R. T. (2021). A new decade and new data at SoyBase, the USDA-ARS soybean genetics and genomics database. *Nucleic Acids Research*, *49*(D1), D1496–D1501.

Bülow, L., Schindler, M., & Hehl, R. (2007). PathoPlant®: A platform for microarray expression data to analyse co-regulated genes involved in plant defense responses. *Nucleic Acids Research*, 35(suppl_1), D841–D845.

Cabrera, J., Bustos, R., Favery, B., Fenoll, C., & Escobar, C. (2014). NEMATIC: A simple and versatile tool for the in silico analysis of plant–nematode interactions. *Molecular Plant Pathology*, *15*(6), 627–636.

Calderone, A., Licata, L., & Cesareni, G. (2015). VirusMentha: A new resource for virus-host protein interactions. *Nucleic Acids Research*, *43*(D1), D588–D592.

Castro-Moretti, F. R., Gentzel, I. N., Mackey, D., & Alonso, A. P. (2020). Metabolomics as an emerging tool for the study of plant – pathogen interactions. *Metabolites*, *10*(2), 52.

Chen, X., Zhang, Y., Zhang, Z., Zhao, Y., Sun, C., Yang, M., Wang, J., Liu, Q., Zhang, B., Chen, M., Yu, J., Wu, J., Jin, Z., & Xiao, J. (2018). PGAweb: A web server for bacterial pan-genome analysis. *Frontiers in Microbiology*, *9*, 1910.

Collmer, A., Schneider, D. J., & Lindeberg, M. (2009). Lifestyles of the effector rich: Genome-enabled characterisation of bacterial plant pathogens. *Plant Physiology*, *150*(4), 1623–1630.

Derevnina, L., Petre, B., Kellner, R., Dagdas, Y. F., Sarowar, M. N., Giannakopoulou, A., Chaparro-Garcia, A., Pennington, H. G., Van West, P., & Kamoun, S. (2016). Emerging oomycete threats to plants and animals. *Philosophical Transactions of the Royal Society B: Biological Sciences*, *371*(1709), 20150459.

Desaint, H., Aoun, N., Deslandes, L., Vailleau, F., Roux, F., & Berthomé, R. (2021). Fight hard or die trying: When plants face pathogens under heat stress. *New Phytologist*, *229*(2), 712–734.

Dong, S., Raffaele, S., & Kamoun, S. (2015a). The two-speed genomes of filamentous pathogens: Waltz with plants. *Current Opinion in Genetics & Development*, *35*, 57–65.

Dong, X., Lu, X., & Zhang, Z. (2015b). BEAN 2.0: An integrated web resource for the identification and functional analysis of type III secreted effectors. *Database*. https://doi.org/10.1093/database/bav064

Du, W., Wang, Y., & Qiao, Y. (2017). *Rpan: An end-to-end recurrent pose-attention network for action recognition in videos* (pp. 3725–3734). Proceedings of the IEEE International Conference on Computer Vision.

Eckardt, N. A. (2000). Sequencing the rice genome. *Plant Cell*, *12*.

Eves-van den Akker, S. (2021). Plant – nematode interactions. *Current Opinion in Plant Biology*, *62*, 102035.

Eves-van den Akker, S., Stojilković, B., & Gheysen, G. (2021). Recent applications of biotechnological approaches to elucidate the biology of plant – nematode interactions. *Current Opinion in Biotechnology*, *70*, 122–130.

Farooq, T., Adeel, M., He, Z., Umar, M., Shakoor, N., da Silva, W., Elmer, W., White, J. C., & Rui, Y. (2021). Nanotechnology and plant viruses: An emerging disease management approach for resistant pathogens. *ACS Nano*, *15*(4), 6030–6037.

Gao, R. F., Wang, J. Y., Liu, K. W., Yoshida, K., Hsiao, Y. Y., Shi, Y. X., Tsai, K. C., Chen, Y. Y., Mitsuda, N., Liang, C. K., Wang, Z. W., Wang, Y., Zhang, D. Y., Huang, L., Zhao, X., Zhong, W. Y., Cheng, Y. H., Jiang, Z. D., . . . Zhang, G. M. (2021). Comparative analysis of Phytophthora genomes reveals oomycete pathogenesis in crops. *Heliyon*, *7*(2), e06317.

Gao, Y., Yang, Z., Yang, W., Yang, Y., Gong, J., Yang, Q. Y., & Niu, X. (2021). PlantImputeDB: An integrated multiple plant reference panel database for genotype imputation. *Nucleic Acids Research*, *49*(D1), D1480–D1488.

Garsmeur, O., Droc, G., Antonise, R., Grimwood, J., Potier, B., Aitken, K., Jenkins, J., Martin, G., Charron, C., Hervouet, C., Costet, L., & D'hont, A. (2018). A mosaic monoploid reference sequence for the highly complex genome of sugarcane. *Nature Communications*, *9*(1), 1–10.

Goffeau, A., Barrell, B. G., Bussey, H., Davis, R. W., Dujon, B., Feldmann, H., Galibert, F., Hoheisel, J. D., Jacq, C., Johnston, M., Louis, E. J., & Oliver, S. G. (1996). Life with 6000 genes. *Science*, *274*(5287), 546–567.

Golicz, A. A., Batley, J., & Edwards, D. (2016). Towards plant pangenomics. *Plant Biotechnology Journal*, *14*(4), 1099–1105.

Goodstein, D. M., Shu, S., Howson, R., Neupane, R., Hayes, R. D., Fazo, J., & Rokhsar, D. S. (2012). Phytozome: A comparative platform for green plant genomics. *Nucleic Acids Research*, *40*(D1), D1178–D1186.

Grigoriev, I. V., Nikitin, R., Haridas, S., Kuo, A., Ohm, R., Otillar, R., & Shabalov, I. (2014). MycoCosm portal: Gearing up for 1000 fungal genomes. *Nucleic Acids Research*, *42*(D1), D699–D704.

Grünwald, N. J., Martin, F. N., Larsen, M. M., Sullivan, C. M., Press, C. M., Coffey, M. D., Hansen, E. M., & Parke, J. L. (2011). Phytophthora-ID. org: A sequence-based Phytophthora identification tool. *Plant Disease*, *95*(3), 337–342.

Hansen, B. O., & Olsson, S. (2020). OmicsDB: Pathogens – A database for exploring functional networks of plant pathogens. *BioRxiv*. https://www.biorxiv.org/content/biorxiv/early/2020/03/19/2020.03.18.979971.full.pdf

Hasan, M. M., Basith, S., Khatun, M. S., Lee, G., Manavalan, B., & Kurata, H. (2021). Meta-i6mA: An interspecies predictor for identifying DNA N 6-methyladenine sites of plant genomes by exploiting informative features in an integrative machine-learning framework. *Briefings in Bioinformatics*, *22*(3), bbaa202.

Hirakawa, H. (2017). Gene annotation and database of the radish. In *The radish genome* (pp. 43–51). Springer.

Hirakawa, H., Shirasawa, K., Miyatake, K. O. J. I., Nunome, T., Negoro, S., Ohyama, A. K. I. O., Yamaguchi, H., Sato, S., Isobe, S., Tabata, S., & Fukuoka, H. (2014). Draft genome sequence of eggplant (Solanum melongena L.): The representative solanum species indigenous to the old world. *DNA Research*, *21*(6), 649–660.

Howe, K. L., Bolt, B. J., Shafie, M., Kersey, P., & Berriman, M. (2017). WormBase Para-Site– a comprehensive resource for helminth genomics. *Molecular and Biochemical Parasitology*, *215*, 2–10.

Hu, T., Chitnis, N., Monos, D., & Dinh, A. (2021). Next-generation sequencing technologies: An overview. *Human Immunology*, *82*.

Ibrahim, H. M., Ahmad, E. M., Martínez-Medina, A., & Aly, M. A. (2019). Effective approaches to study the plant-root knot nematode interaction. *Plant Physiology and Biochemistry*, *141*, 332–342.

Imam, J., Singh, P. K., & Shukla, P. (2016). Plant microbe interactions in post genomic era: Perspectives and applications. *Frontiers in Microbiology*, *7*, 1488.

Iquebal, M. A., Jaiswal, S., Mishra, V. K., Jasrotia, R. S., Angadi, U. B., Singh, B. P., Passari, A. K., Deka, P., Prabha, R., Singh, D. P., Gupta, V. K., & Kumar, D. (2021). Fungal genomic resources for strain identification and diversity analysis of 1900 fungal species. *Journal of Fungi*, *7*(4), 288.

Jones, D. A., Rozano, L., Debler, J., Mancera, R. L., Moolhuijzen, P., & Hane, J. K. (2021). Predector: An automated and combinative method for the predictive ranking of candidate effector proteins of fungal plant-pathogens. *Scientific Reports*, *11*(1), 19731.

Jones, J. T., Haegeman, A., Danchin, E. G., Gaur, H. S., Helder, J., Jones, M. G., & Perry, R. N. (2013). Top 10 plant-parasitic nematodes in molecular plant pathology. *Molecular Plant Pathology*, *14*(9), 946–961.

Juanillas, V., Dereeper, A., Beaume, N., Droc, G., Dizon, J., Mendoza, J. R., & Mauleon, R. P. (2019). Rice galaxy: An open resource for plant science. *Gigascience*, *8*(5), giz028.

Kamoun, S., Furzer, O., Jones, J. D., Judelson, H. S., Ali, G. S., Dalio, R. J., Roy, S. G., Schena, L., Zambounis, A., Panabières, F., Cahill, D., & Govers, F. (2015). The top 10 oomycete pathogens in molecular plant pathology. *Molecular Plant Pathology*, *16*(4), 413–434.

Karasov, T. L., Horton, M. W., & Bergelson, J. (2014). Genomic variability as a driver of plant–pathogen coevolution? *Current Opinion in Plant Biology*, *18*, 24–30.

Karp, P. D., Ivanova, N., Krummenacker, M., Kyrpides, N., Latendresse, M., Midford, P., Ong, W. K., Paley, S., & Seshadri, R. (2019). A comparison of microbial genome web portals. *Frontiers in Microbiology*, *10*, 208.

Kikuchi, T., Eves-van den Akker, S., & Jones, J. T. (2017). Genome evolution of plant-parasitic nematodes. *Annual Review of Phytopathology*, *55*, 333–354.

Kim, S., Park, M., Yeom, S. I., Kim, Y. M., Lee, J. M., Lee, H. A., Seo, E., Choi, J., Cheong, K., Kim, K. T., Jung, K., & Choi, D. (2014). Genome sequence of the hot pepper provides insights into the evolution of pungency in Capsicum species. *Nature Genetics*, *46*(3), 270–278.

Kourelis, J., & Kamoun, S. (2020). RefPlantNLR: A comprehensive collection of experimentally validated plant NLRs. *bioRxiv*. https://doi.org/10.1101/2020.07.08.193961

Kozieł, E., Otulak-Kozieł, K., & Bujarski, J. J. (2021). Plant cell wall as a key player during resistant and susceptible plant-virus interactions. *Frontiers in Microbiology*, *12*, 495.

Kozlowski, D. K., Hassanaly-Goulamhoussen, R., Da Rocha, M., Koutsovoulos, G. D., Bailly-Bechet, M., & Danchin, E. G. (2020). Transposable elements are an evolutionary force shaping genomic plasticity in the parthenogenetic root-knot nematode Meloidogyne incognita. *BioRxiv*, *4*.

Kumar, G., & Nanda, S. (2021). Molecular perspectives of plant-pathogen interactions: An overview on plant immunity. *Biological Forum – An International Journal*, *13*(1), 48–53.

Kushwaha, S. K., Chauhan, P., Hedlund, K., & Ahrén, D. (2016). NBSPred: A support vector machine-based high-throughput pipeline for plant resistance protein NBSLRR prediction. *Bioinformatics*, *32*(8), 1223–1225.

Kutnjak, D., Tamisier, L., Adams, I., Boonham, N., Candresse, T., Chiumenti, M., De Jonghe, K., Kreuze, J. F., Lefebvre, M., Silva, G., Malapi-Wight, M., Margaria, P., Pleško, I. M., McGreig, S., Miozzi, L., Remenant, B., Reynard, J. S., Rollin, J., Rott, M., . . . Haegeman, A. (2021). A primer on the analysis of high-throughput sequencing data for detection of plant viruses. *Microorganisms*, *9*(4), 841.

Langeland, A., Hawdon, J. M., & O'Halloran, D. M. (2021). NemChR-DB: A database of parasitic nematode chemosensory G-protein coupled receptors. *International Journal for Parasitology*, *51*(5), 333–337.

Lefeuvre, P., Martin, D. P., Elena, S. F., Shepherd, D. N., Roumagnac, P., & Varsani, A. (2019). Evolution and ecology of plant viruses. *Nature Reviews Microbiology*, *17*(10), 632–644.

Li, F. W., & Harkess, A. (2018). A guide to sequence your favorite plant genomes. *Applications in Plant Sciences*, *6*(3).

Li, F., Fan, G., Wang, K., Sun, F., Yuan, Y., Song, G., Li, Q., Ma, Z., Lu, C., Zou, C., Chen, W., & Yu, S. (2014a). Genome sequence of the cultivated cotton *Gossypium arboreum*. *Nature Genetics*, *46*(6), 567–572.

Li, H., & Zhang, Z. (2016). Systems understanding of plant – pathogen interactions through genome-wide protein – protein interaction networks. *Frontiers of Agricultural Science and Engineering*, *3*(2), 102–112.

Li, J. R., Liu, C. C., Sun, C. H., & Chen, Y. T. (2018). Plant stress RNA-seq Nexus: A stress-specific transcriptome database in plant cells. *BMC Genomics*, *19*(1), 1–8.

Li, J. Y., Wang, J., & Zeigler, R. S. (2014b). The 3,000 rice genomes project: New opportunities and challenges for future rice research. *Gigascience*, *3*(1), 2047–217X.

Li, J., Tai, C., Deng, Z., Zhong, W., He, Y., & Ou, H. Y. (2018). VRprofile: Gene-cluster-detection-based profiling of virulence and antibiotic resistance traits encoded within genome sequences of pathogenic bacteria. *Briefings in Bioinformatics*, *19*(4), 566–574.

Li, P., Quan, X., Jia, G., Xiao, J., Cloutier, S., & You, F. M. (2016). RGAugury: A pipeline for genome-wide prediction of resistance gene analogs (RGAs) in plants. *BMC Genomics*, *17*(1), 1–10.

Li, Y., Steenwyk, J. L., Chang, Y., Wang, Y., James, T. Y., Stajich, J. E., Spatafora, J.W., Groenewald, M., Dunn, C. W., Hittinger, C. T., Shen, X. X., & Rokas, A. (2021). A genome-scale phylogeny of the kingdom Fungi. *Current Biology*, *31*(8), 1653–1665.

Ling, H. Q., Zhao, S., Liu, D., Wang, J., Sun, H., Zhang, C., Fan, H., Li, D., Dong, L., Tao, Y., Gao, C., & Wang, J. (2013). Draft genome of the wheat A-genome progenitor Triticum urartu. *Nature*, *496*(7443), 87–90.

Liu, B., Zheng, D., Jin, Q., Chen, L., & Yang, J. (2019). VFDB 2019: A comparative pathogenomic platform with an interactive web interface. *Nucleic Acids Research*, *47*(D1), D687–D692.

Liu, M., Li, Y., Ma, Y., Zhao, Q., Stiller, J., Feng, Q., Tian, Q., Liu, D., Han, B., & Liu, C. (2020). The draft genome of a wild barley genotype reveals its enrichment in genes related to biotic and abiotic stresses compared to cultivated barley. *Plant Biotechnology Journal*, *18*(2), 443–456.

Loaiza, C. D., & Kaundal, R. (2021). PredHPI: An integrated web server platform for the detection and visualisation of host–pathogen interactions using sequence-based methods. *Bioinformatics*, *37*(5), 622–624.

Lücking, R., Aime, M. C., Robbertse, B., Miller, A. N., Aoki, T., Ariyawansa, H. A., Cardinali, G., Crous, P. W., Druzhinina, I. S., Geiser, D. M., Hawksworth, D. L., Hyde, K. D., Irinyi, L., Jeewon, R., Johnston, P. R., Kirk, P. M., Malosso, E., May, T. M., Meyer, W., . . . Schoch, C. L. (2021). Fungal taxonomy and sequence-based nomenclature. *Nature Microbiology*, *6*(5), 540–548.

Madronero, L. J., Corredor-Rozo, Z. L., Escobar-Perez, J., & Velandia-Romero, M. L. (2019). Next generation sequencing and proteomics in plant virology: How is Colombia doing? *Acta Biológica Colombiana*, *24*(3), 423–438.

Marthey, S., Aguileta, G., Rodolphe, F., Gendrault, A., Giraud, T., Fournier, E., Lopez-Villavicencio, M., Gautier, A., Lebrun, M. H., & Chiapello, H. (2008). FUNYBASE: A FUNgal phYlogenomic dataBASE. *BMC Bioinformatics*, *9*(1), 1–10.

Martin, J., Rosa, B. A., Ozersky, P., Hallsworth-Pepin, K., Zhang, X., Bhonagiri-Palsikar, V., Tyagi, R., Wang, Q., Choi, Y. J., Gao, X., McNulty, S. N., & Mitreva, M. (2015). Helminth. net: Expansions to Nematode. net and an introduction to Trematode. net. *Nucleic Acids Research*, *43*(D1), D698–D706.

Masonbrink, R., Maier, T. R., Seetharam, A. S., Juvale, P. S., Baber, L., Baum, T. J., & Severin, A. J. (2019). SCNBase: A genomics portal for the soybean cyst nematode (Heterodera glycines). *Database*. https://doi.org/10.1093/database/baz111

McCarthy, C. G., & Fitzpatrick, D. A. (2017). Phylogenomic reconstruction of the oomycete phylogeny derived from 37 genomes. *Msphere*, *2*(2).

McGowan, J., Byrne, K. P., & Fitzpatrick, D. A. (2019). Comparative analysis of oomycete genome evolution using the oomycete gene order browser (OGOB). *Genome Biology and Evolution*, *11*(1), 189–206.

McGowan, J., & Fitzpatrick, D. A. (2020). Recent advances in oomycete genomics. *Advances in Genetics*, *105*, 175–228.

Mihara, T., Nishimura, Y., Shimizu, Y., Nishiyama, H., Yoshikawa, G., Uehara, H., Hingamp, P., Goto, S., & Ogata, H. (2016). Linking virus genomes with host taxonomy. *Viruses*, *8*(3), 66.

Möller, M., & Stukenbrock, E. H. (2017). Evolution and genome architecture in fungal plant pathogens. *Nature Reviews Microbiology*, *15*(12), 756.

Montarry, J., Mimee, B., Danchin, E. G., Koutsovoulos, G. D., Ste-Croix, D. T., & Grenier, E. (2021). Recent advances in population genomics of plant-parasitic nematodes. *Phytopathology®*, *111*(1), 40–48.

Mukherjee, S., Stamatis, D., Bertsch, J., Ovchinnikova, G., Sundaramurthi, J. C., Lee, J., & Reddy, T. B. K. (2021). Genomes online database (GOLD) v. 8: Overview and updates. *Nucleic Acids Research*, *49*(D1), D723–D733.

Naidoo, S., Visser, E. A., Zwart, L., Du Toit, Y., Bhadauria, V., & Shuey, L. S. (2017). Dual RNA-seq to elucidate the plant – pathogen duel. *Current Issues in Molecular Biology*, *27*(1), 127–142.

Nakaya, A., Ichihara, H., Asamizu, E., Shirasawa, S., Nakamura, Y., Tabata, S., & Hirakawa, H. (2017). Plant genome database Japan (PGDBj). In *Plant genomics databases* (pp. 45–77). Humana Press.

Nur, M., Wood, K., & Michelmore, R. (2021). EffectorO: Motif-independent prediction of effectors in oomycete genomes using machine learning and lineage specificity. *bioRxiv*. https://doi.org/10.1101/2021.03.19.436227

Opperman, C. H., Bird, D. M., Williamson, V. M., Rokhsar, D. S., Burke, M., Cohn, J., Cromer, J., Diener, S., Gajan, J., Graham, S., Houfek, T. D., & Windham, E. (2008). Sequence and genetic map of Meloidogyne hapla: A compact nematode genome for plant parasitism. *Proceedings of the National Academy of Sciences*, *105*(39), 14802–14807.

Osuna-Cruz, C. M., Paytuvi-Gallart, A., Di Donato, A., Sundesha, V., Andolfo, G., Aiese Cigliano, R., & Ercolano, M. R. (2018). PRGdb 3.0: A comprehensive platform for prediction and analysis of plant disease resistance genes. *Nucleic Acids Research*, *46*(D1), D1197–D1201.

Pal, T., Jaiswal, V., & Chauhan, R. S. (2016). DRPPP: A machine learning based tool for prediction of disease resistance proteins in plants. *Computers in Biology and Medicine*, *78*, 42–48.

Parrello, B., Butler, R., Chlenski, P., Pusch, G. D., & Overbeek, R. (2021). Supervised extraction of near-complete genomes from metagenomic samples: A new service in PATRIC. *PLoS One*, *16*(4), e0250092.

Pedro, H., Maheswari, U., Urban, M., Irvine, A. G., Cuzick, A., McDowall, M. D., Staines, D. M., Kulesha, E., Hammond-Kosack, K. E., & Kersey, P. J. (2016). PhytoPath: An integrative resource for plant pathogen genomics. *Nucleic Acids Research*, *44*(D1), D688–D693.

Peyraud, R., Dubiella, U., Barbacci, A., Genin, S., Raffaele, S., & Roby, D. (2017). Advances on plant – pathogen interactions from molecular toward systems biology perspectives. *The Plant Journal*, *90*(4), 720–737.

Phan, N. T., Orjuela, J., Danchin, E. G., Klopp, C., Perfus-Barbeoch, L., Kozlowski, D. K., Koutsovoulos, G. D., Lopez-Roques, C., Bouchez, O., Zahm, M., Besnard, G., & Bellafiore, S. (2020). Genome structure and content of the rice root-knot nematode (Meloidogyne graminicola). *Ecology and Evolution*, *10*(20), 11006–11021.

Phani, V., Khan, M. R., & Dutta, T. K. (2021). Plant-parasitic nematodes as a potential threat to protected agriculture: Current status and management options. *Crop Protection*, 105573.

Portwood, J. L., Woodhouse, M. R., Cannon, E. K., Gardiner, J. M., Harper, L. C., Schaeffer, M. L., Walsh, J. R., Sen, T. Z., Cho, K. T., Schott, D. A., Braun, B. L., & Andorf, C. M. (2019). MaizeGDB 2018: The maize multi-genome genetics and genomics database. *Nucleic Acids Research*, *47*(D1), D1146–D1154.

Potato Genome Sequencing Consortium. (2011). Genome sequence and analysis of the tuber crop potato. *Nature*, *475*(7355), 189.

Powers, T. O., Harris, T. S., Higgins, R. S., Mullin, P. G., & Powers, K. S. (2021). Nematode biodiversity assessments need vouchered databases: A BOLD reference library for plant-parasitic nematodes in the superfamily Criconematoidea. *Genome*, *64*(3), 232–241.

Qi, H., Jiang, Z., Zhang, K., Yang, S., He, F., & Zhang, Z. (2018). PlaD: A transcriptomics database for plant defense responses to pathogens, providing new insights into plant immune system. *Genomics, Proteomics & Bioinformatics*, *16*(4), 283–293.

Qing, X., Wang, M., Karssen, G., Bucki, P., Bert, W., & Braun-Miyara, S. (2020). PPNID: A reference database and molecular identification pipeline for plant-parasitic nematodes. *Bioinformatics*, *36*(4), 1052–1056.

Riaño-Pachón, D. M., & Mattiello, L. (2017). Draft genome sequencing of the sugarcane hybrid SP80–3280. *F1000Research*, *6*.

Rujirawat, T., Patumcharoenpol, P., Kittichotirat, W., & Krajaejun, T. (2019). Oomycete gene table: An online database for comparative genomic analyses of the oomycete microorganisms. *Database*. https://doi.org/10.1093/database/baz082

Sabbagh, C. R. R., Carrere, S., Lonjon, F., Vailleau, F., Macho, A. P., Genin, S., & Peeters, N. (2019). Pangenomic type III effector database of the plant pathogenic Ralstonia spp. *PeerJ*, *7*, e7346.

Schneider, K., Van der Werf, W., Cendoya, M., Mourits, M., Navas-Cortés, J. A., Vicent, A., & Lansink, A. O. (2020). Impact of Xylella fastidiosa subspecies pauca in European olives. *Proceedings of the National Academy of Sciences*, *117*(17), 9250–9259.

Shirasawa, K., Harada, D., Hirakawa, H., Isobe, S., & Kole, C. (2021). Chromosome-level de novo genome assemblies of over 100 plant species. *Breeding Science*, 20146.

Sierro, N., Battey, J. N., Ouadi, S., Bakaher, N., Bovet, L., Willig, A., Goepfert, S., Peitsch, M. C., & Ivanov, N. V. (2014). The tobacco genome sequence and its comparison with those of tomato and potato. *Nature Communications*, *5*(1), 1–9.

Silva, R. J. S., & Micheli, F. (2020). RRGPredictor, a set-theory-based tool for predicting pathogen-associated molecular pattern receptors (PRRs) and resistance (R) proteins from plants. *Genomics*, *112*(3), 2666–2676.

Simpson, A. J. G., Reinach, F. D. C., Arruda, P., Abreu, F. A. D., Acencio, M., Alvarenga, R., Alves, L. C., Araya, J. E., Baia, G. S., Baptista, C. S., Barros, M. H. D., & Setubal, J. C. (2000). The genome sequence of the plant pathogen Xylella fastidiosa. *Nature*, *406*(6792), 151–157.

Souza, P. F., Garcia-Ruiz, H., & Carvalho, F. E. (2019). What proteomics can reveal about plant – virus interactions? Photosynthesis-related proteins on the spotlight. *Theoretical and Experimental Plant Physiology*, *31*(1), 227–248.

Spannagl, M., Nussbaumer, T., Bader, K. C., Martis, M. M., Seidel, M., Kugler, K. G., Gundlach, H., & Mayer, K. F. (2016). PGSB PlantsDB: Updates to the database framework for comparative plant genome research. *Nucleic Acids Research*, *44*(D1), D1141–D1147.

Sperschneider, J., Catanzariti, A. M., DeBoer, K., Petre, B., Gardiner, D. M., Singh, K. B., Dodds, P. N., & Taylor, J. M. (2017). LOCALISER: Subcellular localisation prediction of both plant and effector proteins in the plant cell. *Scientific Reports*, *7*(1), 1–14.

Sperschneider, J., Dodds, P. N., Gardiner, D. M., Singh, K. B., & Taylor, J. M. (2018a). Improved prediction of fungal effector proteins from secretomes with EffectorP 2.0. *Molecular Plant Pathology*, *19*(9), 2094–2110.

Sperschneider, J., Dodds, P. N., Singh, K. B., & Taylor, J. M. (2018b). ApoplastP: Prediction of effectors and plant proteins in the apoplast using machine learning. *New Phytologist*, *217*(4), 1764–1778.

Steuernagel, B., Witek, K., Krattinger, S. G., Ramirez-Gonzalez, R. H., Schoonbeek, H. J., Yu, G., Baggs, E., Witek, A. I., Yadav, I., Krasileva, K. V., Jones, J. D., & Wulff, B. B. (2020). The NLR-Annotator tool enables annotation of the intracellular immune receptor repertoire. *Plant Physiology*, *183*(2), 468–482.

Straub, C., Colombi, E., & McCann, H. C. (2021). Population genomics of bacterial plant pathogens. *Phytopathology®*, *111*(1), 23–31.

Toda, N., Rustenholz, C., Baud, A., Le Paslier, M. C., Amselem, J., Merdinoglu, D., & Faivre-Rampant, P. (2020). NLGenomeSweeper: A tool for genome-wide NBS-LRR resistance gene identification. *Genes*, *11*(3), 333.

Tomato Genome Consortium. (2012). The tomato genome sequence provides insights into fleshy fruit evolution. *Nature*, *485*(7400), 635.

Tripathy, S., Deo, T., & Tyler, B. M. (2012). Oomycete transcriptomics database: A resource for oomycete transcriptomes. *BMC Genomics*, *13*(1), 1–12.

Trivellone, V. (2019). An online global database of hemiptera-phytoplasma-plant biological interactions. *Biodiversity Data Journal*, *7*.

Uchiyama, I., Mihara, M., Nishide, H., Chiba, H., & Kato, M. (2019). MBGD update 2018: Microbial genome database based on hierarchical orthology relations covering closely related and distantly related comparisons. *Nucleic Acids Research*, *47*(D1), D382–D389.

Urban, M., Cuzick, A., Seager, J., Wood, V., Rutherford, K., Venkatesh, S. Y., De Silva, N., Martinez, M. C., Pedro, H., Yates, A. D., Hassani-Pak, K., & Hammond-Kosack, K. E. (2020). PHI-base: The pathogen – host interactions database. *Nucleic Acids Research*, *48*(D1), D613–D620.

Valentin, G., Abdel, T., Gaëtan, D., Jean-François, D., Matthieu, C., & Mathieu, R. (2021). GreenPhylDB v5: A comparative pangenomic database for plant genomes. *Nucleic Acids Research*, *49*(D1), D1464–D1471.

van den Hoogen, J., Geisen, S., Wall, D. H., Wardle, D. A., Traunspurger, W., de Goede, R. G., Adams, B.J., Ahmad, W., Ferris, H., Bardgett, R. D., Bonkowski, M., & Crowther, T. W. (2020). A global database of soil nematode abundance and functional group composition. *Scientific Data*, *7*(1), 1–8.

Vasar, M., Davison, J., Neuenkamp, L., Sepp, S. K., Young, J. P. W., Moora, M., & Öpik, M. (2021). User-friendly bioinformatics pipeline gDAT (graphical downstream analysis tool) for analysing rDNA sequences. *Molecular Ecology Resources*, *21*(4), 1380–1392.

Větrovský, T., Morais, D., Kohout, P., Lepinay, C., Algora, C., Hollá, S. A., Bahnmann, B. D., Bílohnědá, K., Brabcová, V., D'Alò, F., Human, Z. R., & Baldrian, P. (2020). GlobalFungi, a global database of fungal occurrences from high-throughput-sequencing metabarcoding studies. *Scientific Data*, *7*(1), 1–14.

Vinatzer, B. A., & Yan, S. (2008). Mining the genomes of plant pathogenic bacteria: How not to drown in gigabases of sequence. *Molecular Plant Pathology*, *9*(1), 105–118.

Walkowiak, S., Gao, L., Monat, C., Haberer, G., Kassa, M. T., Brinton, J., Ramirez-Gonzalez, R. H., Kolodziej, M. C., Delorean, E., Thambugala, D., Klymiuk, V., & Pozniak, C. J. (2020). Multiple wheat genomes reveal global variation in modern breeding. *Nature*, 1–7.

Wan, Y., Renner, D. W., Albert, I., & Szpara, M. L. (2015). VirAmp: A galaxy-based viral genome assembly pipeline. *Gigascience*, *4*(1), s13742–015.

Wang, C., Wang, P., Han, S., Wang, L., Zhao, Y., & Juan, L. (2020). FunEffector-Pred: Identification of fungi effector by activate learning and genetic algorithm sampling of imbalanced data. *IEEE Access*, *8*, 57674–57683.

Wang, Y., Wang, P., Guo, Y., Huang, S., Chen, Y., & Xu, L. (2021). prPred: A predictor to identify plant resistance proteins by incorporating k-spaced amino acid (Group) Pairs. *Frontiers in Bioengineering and Biotechnology*, *8*, 1593.

Wattam, A. R., Abraham, D., Dalay, O., Disz, T. L., Driscoll, T., Gabbard, J. L., Gillespie, J. J., Gough, R., Hix, D., Kenyon, R., Machi, D., & Sobral, B. W. (2014). PATRIC, the bacterial bioinformatics database and analysis resource. *Nucleic Acids Research*, *42*(D1), D581–D591.

Westermann, A. J., Gorski, S. A., & Vogel, J. (2012). Dual RNA-seq of pathogen and host. *Nature Reviews Microbiology*, *10*(9), 618–630.

Xu, J. R., Peng, Y. L., Dickman, M. B., & Sharon, A. (2006). The dawn of fungal pathogen genomics. *Annual Review of Phytopathology*, *44*, 337–366.

Xu, J. R., & Wang, N. (2019). Where are we going with genomics in plant pathogenic bacteria? *Genomics*, *111*(4), 729–736.

Yagi, M. (2018). Recent progress in whole genome sequencing, high-density linkage maps, and genomic databases of ornamental plants. *Breeding Science*, *68*, 17080.

Yang, S., Li, H., He, H., Zhou, Y., & Zhang, Z. (2019). Critical assessment and performance improvement of plant – pathogen protein – protein interaction prediction methods. *Briefings in Bioinformatics*, *20*(1), 274–287.

Zarattini, M., Farjad, M., Launay, A., Cannella, D., Soulié, M. C., Bernacchia, G., & Fagard, M. (2021). Every cloud has a silver lining: How abiotic stresses affect gene expression in plant-pathogen interactions. *Journal of Experimental Botany*, *72*(4), 1020–1033.

Zheng, Y., Gao, S., Padmanabhan, C., Li, R., Galvez, M., Gutierrez, D., Fuentes, S., Ling, K. S., Kreuze, J., & Fei, Z. (2017). VirusDetect: An automated pipeline for efficient virus discovery using deep sequencing of small RNAs. *Virology*, *500*, 130–138.

Zhulin, I. B. (2015). Databases for microbiologists. *Journal of Bacteriology*, *197*(15), 2458–2467.

3 Exploring the Herbicide Resistance Mechanisms in Weeds Using Omics Technologies

Madhab Kumar Sen, Kateřina Hamouzová, Pavlína Košnarová, Josef Soukup, and Amit Roy

3.1 INTRODUCTION

The modern agriculture system aims to reduce economic losses in agricultural production. The reciprocal relationship between the growing population and decreasing area suitable for agriculture will further complicate this scenario. Consequently, the focus has bowed to increased production without increased area under agriculture. Although several factors are involved in gross agricultural economic loss, weeds are considered the most critical biotic factors to reduce crop yields significantly. Some weeds, such as *Apera spica-venti (L.) P. Beauv, Bromus sterilis L., Echinochloa spp.*, blackgrass, barnyardgrass, *Lolium spp.* etc., if left uncontrolled, can result in economic loss that surpasses 50–60% or even 100% of the marketable yield (Gharde et al., 2018). According to the Weed Science Society of America, weeds are defined as "plants growing where it is not desired". Weeds compete with the major crops for resources such as water, light and nutrients etc. and decrease the yield and productivity of the major crops (Baucom, 2019). Herbicides have been widely used to control weeds and improve crop yield quality and quantity (Sharma et al., 2019). The first commercialized herbicides were 2,4-Dichlorophenoxyacetic acid (2,4-D) and 4-Chloro-2-methylphenoxyacetic acid (MCPA). Phenoxyacetic acids, including 2,4-D and various analogues, were widely used during World War II and were further developed after the war (Reade & Cobb, 2002). The mode of action of currently used herbicides was introduced between 1955 and 1975. Since the 1980s, only three new herbicide modes of action have been introduced (Reade & Cobb, 2002). In past decades, herbicides were extensively used to improve crop yield and quality to meet increasing demand. However, uncontrolled use of herbicides with parallel modes of action has resulted in the evolution of herbicide-resistant weeds.

Common herbicide modes of action include interference with photosynthesis (triazines, pyridazinones, amides, nitriles etc.), inhibition of photosynthetic

DOI: 10.1201/9781003153481-3

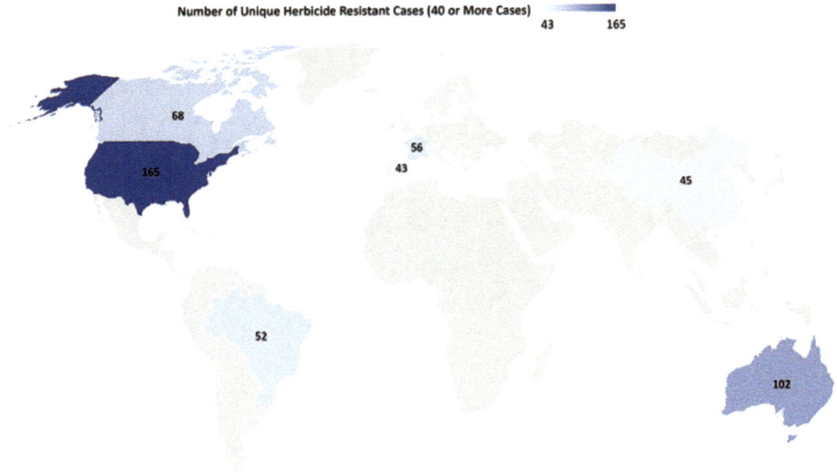

FIGURE 3.1 The number of unique herbicide resistance cases (as of 2020). Countries with 40 or more unique herbicide-resistant cases are included in this figure.

Source: www.weedscience.org/

pigment biosynthesis (triazolinones, oxadizole, thiadiazoles etc.), inhibition of lipid biosynthesis (aryloxyphenoxypropionates, cyclohexanediones and pinoxaden etc.), inhibition of amino acid biosynthesis (sulfonylureas, triazolopyrimidines etc.), inhibition of cellulose biosynthesis (nitriles, benzamides etc.), microtubule disruptors and the cell division inhibitors (benzoic acid inhibitors, carbamates, acetamides etc.) and others. Herbicide resistance refers to the natural ability of a weedy plant to survive a dose that would usually kill an individual of that species (Moss, 2002). The terms "cross-resistance" and "multiple resistance" are often used to describe weeds developing resistance against more than one herbicide due to the same and different modes of detoxification, respectively. However, it should be kept in mind that no universally accepted definitions of "cross-resistance" and "multiple-resistance" exist (Hall et al., 2018). Figure 3.1 represents the number of unique herbicide resistance cases, and Figure 3.2A denotes the number of resistant weed species reported for each herbicide group. In addition, the evolution of multiple herbicide resistance has also been reported in many weed species like *Apera spica venti*, *Lolium rigidum*, *Eleusine indica* etc. (Košnarová et al., 2021; Owen et al., 2007; Vazquez-Garcia et al., 2021; Figure 3.2B).

The mechanisms of herbicide resistance can be divided into target and non-target sites mechanisms (Figure 3.3). Target-site resistance mechanisms (TSR) involve a change to the protein that binds the herbicide, resulting in a lack of inhibition of the biochemical pathway. The most obvious of these is where a mutation within the target protein reduces or eliminates the binding of the herbicide (Sen et al., 2021; Rey-Caballero et al., 2017). Although most of these mutations occur around the herbicide binding site of the target enzyme, some mutations appear elsewhere within the protein structure. In addition to the nucleotide

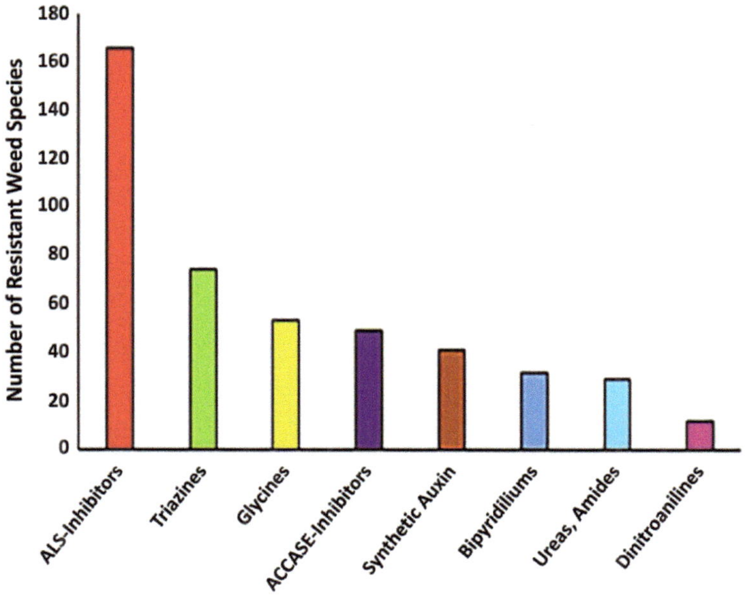

A.

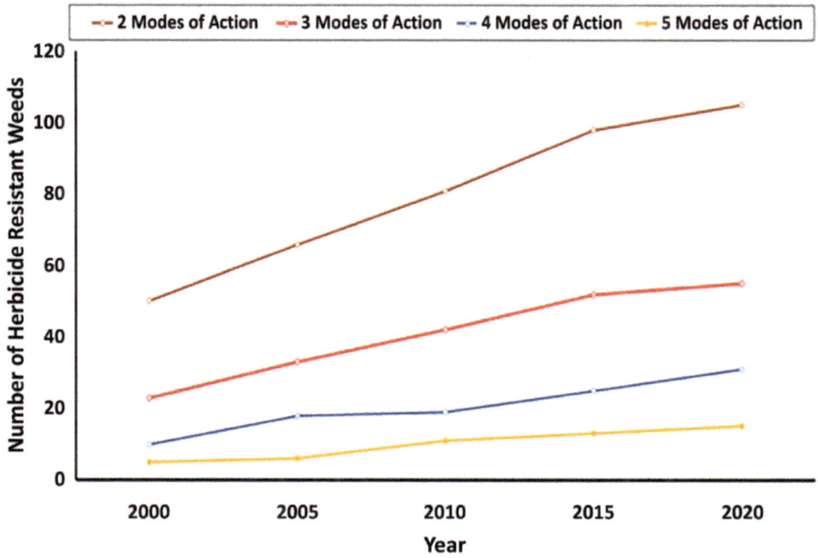

B.

FIGURE 3.2 Status of the number of herbicide-resistant weeds. A. Number of resistant weed species reported for each herbicide group globally (as of 2020), B. number of multiple herbicide-resistance weeds, year-wise.

Source: www.weedscience.org/

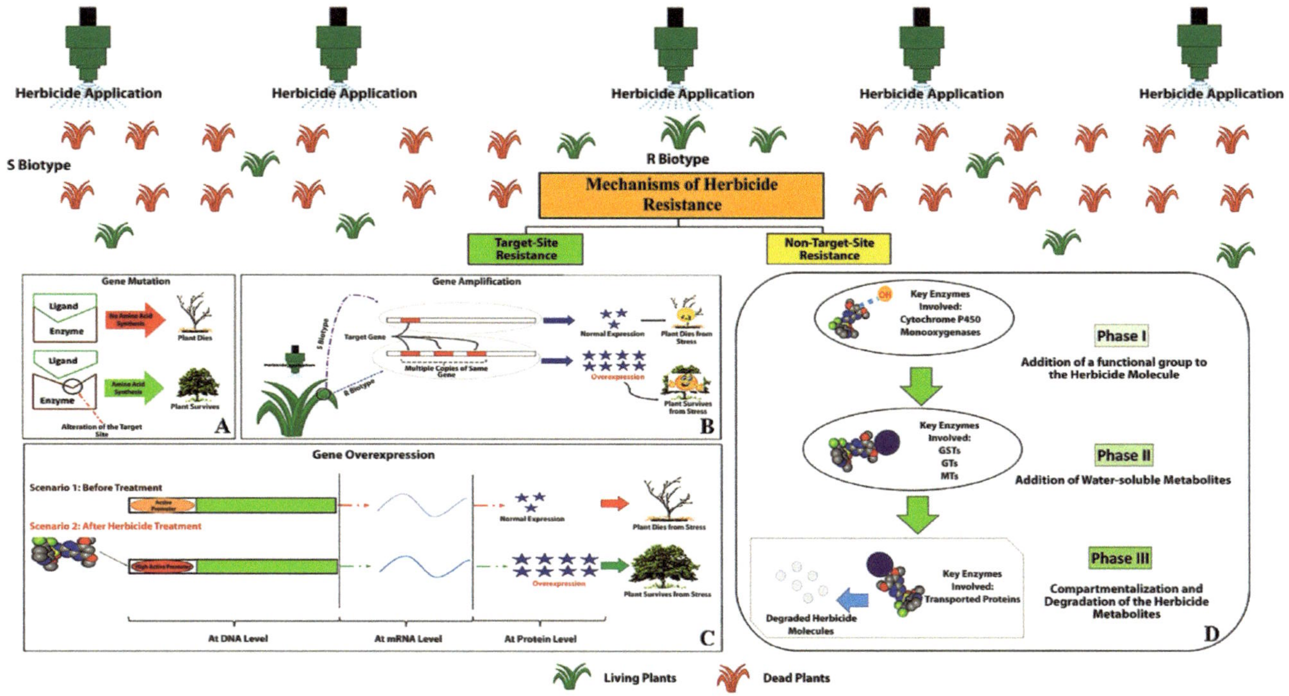

FIGURE 3.3 A model of the mechanisms of herbicide resistance. A. Mutations within the target protein alter the binding site of the herbicide, as a result of which the herbicide could not bind with its target enzyme, thereby conferring resistance in R plants; B. Amplification of the coding region of the target gene results in additional copies within the genome of an organism, thus leading to increased gene expression. Gene amplification resulting in enhanced expression at the mRNA level contributes to the survival of the R plants, following the herbicide application; C. Overexpression of the coding region of the herbicide target gene resulting in the over-production of the target enzyme thus, contributing to resistance in R plants; D. Non-target site mechanisms include herbicide detoxification. Herbicide detoxification is mainly divided into three phases: addition of a functional group to the herbicide molecule (phase I), the addition of water-soluble metabolites (phase II) and, finally, compartmentalization and degradation of the herbicide metabolites (phase III). R: resistant; S: susceptible; GST: Glutathione S-transferase; GT: glucosyltransferases; MT: malonyl transferase.

polymorphisms, target-site mutations might involve whole-codon deletions too. Whole-codon deletions are also known to reduce herbicide's affinity towards the target-site enzyme. Even though these single nucleotide polymorphisms (SNPs) are known to reduce the normal enzymatic functioning, reports suggest that they can retain the full enzymatic functioning (Gaines et al., 2020). The first discovered SNP related to the herbicide resistance was related to the D1 protein encoded by the *psbA* gene. Several known mutation points were reported for herbicide target enzymes like Acetyl-CoA carboxylase (ACCase; EC 6.4.1.2), acetolactate synthase (ALS; EC 2.2.1.6) and as aceto-hydroxy acid synthase (AHAS) etc. To date, codon deletions resulting in herbicide resistance have been reported only in the case of protoporphyrinogen oxidase (PPO; EC 1.3.3.4) in *Amaranthus sp.* The codon deletion removes a glycine residue at the 210th position of PPO (Patzoldt et al., 2006). Apart from single and multiple nucleotide mutations, TSR can also develop due to increased expression of the target-site gene. The overexpression of the target-site gene can be due either to changes in the regulatory regions or increased genomic copy numbers of the particular target gene. Gene duplication results in an extra copy of the coding segment of DNA, which causes increased gene expression at the mRNA levels. Mutations might also accumulate in these new copies, which might cause gain in function or loss of function over time. Evidence for elevated expression of the target gene has been reported in several species like *Bromus sterilis* L. (Sen et al., 2021), *Lolium rigidum* (Baerson et al., 2002), *Conyza canadensis* (Gaines et al., 2020), *Conyza bonariensis* (Gaines et al., 2020), *Hordeum leporinum* (Yu et al., 2007) etc.

Non-target site resistance mechanisms (NTSR) allow plants to survive the herbicide application by not allowing sufficient herbicide to reach the target site. The weed may be initially affected by the herbicide application but eventually will survive and produce seeds. One outcome of this type of mechanism is that it frequently leads to cross-resistance to herbicides of different modes of action. Such phenomena severely complicate resistance management with herbicides (Yuan et al., 2007; Délye et al., 2011, 2013). Cross-resistance patterns tend to be highly variable and unpredictable. The most common NTSR mechanisms are enhanced metabolism, increased detoxification, reduced absorption, reduced translocation and vacuolar sequestration (Sen et al., 2021). In general, herbicide detoxification can be divided into three phases: Phase I (addition of a functional group to the herbicide; mediated by cytochrome P450 monooxygenases), Phase II (addition of water-soluble metabolites; mediated by GSH S-transferases (GST) and glucosyltransferases (GT)] and Phase III (compartmentalization of the herbicide metabolites; mediated by ABC-transporters and other transporters) (Ghanizadeh & Harrington, 2017; Gaines et al., 2020). The evolution of rapid detoxification has been reported in many weeds like *Lolium rigidum* (Christopher et al., 1991) and *Alopecurus myosuroides* Huds (Hall et al., 1995). Reduced absorption across the cuticle and reduced translocation affecting the proper functioning of these systemic herbicides has been reported in many weedy species like *Lolium rigidum* (Lorraine-Colwill et al., 2002; Yu et al., 2004), *Lolium multiflorum* (Vila-Aiub et al., 2013) etc.

High throughput molecular biology-based techniques such as transcriptomics, epitranscriptomics, proteomics and metabolomics and their integrations have the potential to understand better how weeds interact with stresses and develop resistance to herbicides (Figure 3.4). Significant developments have been achieved in model crop species, where the omics-based approaches have uplifted our levels of understanding and application. However, the question remains – how can these omics-based approaches contribute to our knowledge of herbicide resistance and eventually help us develop novel weed management strategies? Genomics and transcriptomics have been the most utilized among the various omics techniques applied within weed science. Recently, in the last few years, metabolomics has also been used frequently. Therefore, proteomics, metabolomics and epigenetics can be deemed emerging prospective research areas for herbicide resistance. Hence, this chapter will provide a detailed outline of how the existing omics technologies can be used to unscramble the novel mechanisms of herbicide target sites, which will be critical to decoding how the new herbicides may exert their action.

3.2 DNA-BASED OMICS APPROACHES

3.2.1 Genomics in Herbicide Resistance

Even though the construction of reference genome assemblies has enhanced the study of basic biology and evolution in many species, these resources have yet to be developed for many agronomically important weeds. Currently, there are only eight weeds with significant research efforts. Unfortunately, only three of these eight (*Oryza ssp.*, *Helianthus ssp.* and *Lolium perenne*) have close crop relatives. Others do not have any close crop relatives (Horvath, 2010). To date, waterhemp (*Amaranthus tuberculatus*) is the only weed to have significant genomic sequence data published (Horvath, 2010; Montgomery et al., 2020). Modern genetics heavily counts in introducing foreign gene/s into the genome (or transformation). The construction of mutant libraries will enable the weed molecular scientists to get large and diversely complexed proteins that can be used to detect the critical residues and optimize protein function. Besides detecting the structurally and functionally critical residues, mutant libraries will enable us to obtain novel proteins with enhanced functions such as increased expression levels, stability, enzymatic activity etc. (You & Zhang, 2012; Abou-Nader & Benedik, 2010). Although mutant libraries help identify several exciting questions related to evolution, the development of successful mutation libraries relies heavily upon transformation systems. Very few weeds have been transformed, limiting the confirmation of the hypotheses derived from weed functional genomic experiments. Hence, as of yet, no mutant libraries are developed for weedy species. Several fundamental questions, such as the characteristics of weeds, genes and physiological processes required for developing invasiveness and weediness, can be answered by introducing the genomic-based tools and technologies in weed science (Ravet et al., 2018). Despite having minimal resources for genomics studies in weedy

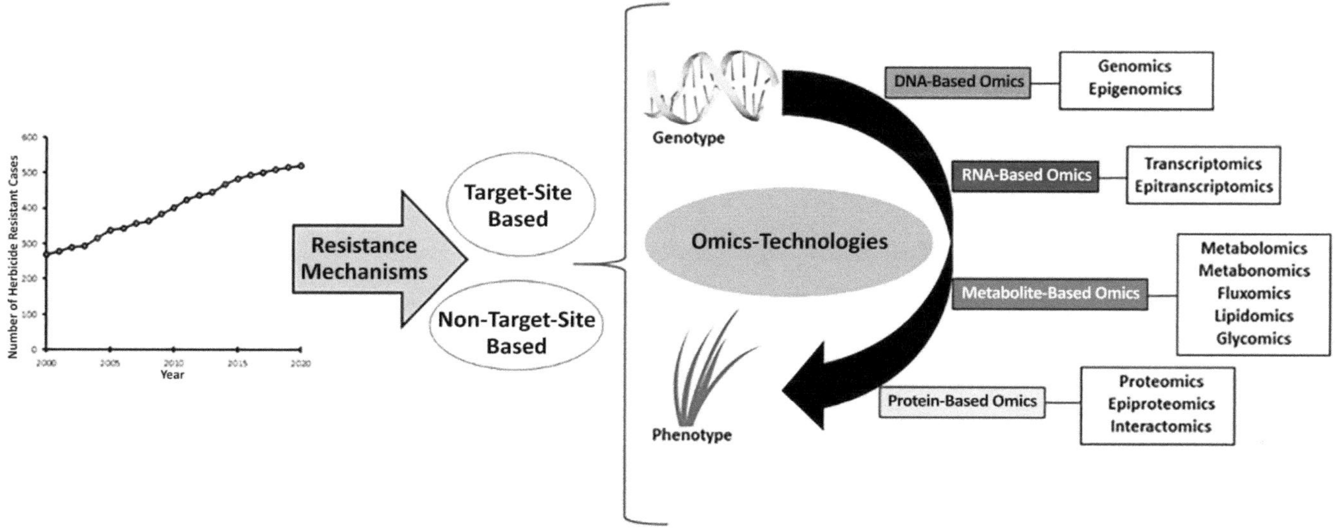

FIGURE 3.4 Potential applications of omics technologies in unravelling the mechanisms of herbicide resistance development. Omics methods will aid a greater understanding of herbicide resistance mechanisms at each level of the central dogma of molecular biology. Such understanding will facilitate future formulations for superior resistance management.

plant species, the genomic resources established for model plants and crops such as *Arabidopsis sp.*, *Zea mays*, *Oryza sp.*, *Solanum tuberosum* can be utilized to clone candidate genes from the non-model species such as weeds. The target sites for ALS inhibitors were initially cloned from *Arabidopsis* (Haughn et al., 1988). After that, it became possible to use the sequence data from the model species and crops to explore the target sites of various herbicides in weeds due to the conservative nature of many genes (Horvath, 2010).

To date, single-gene sequencing and identifying single-point mutations within the gene of interest have mainly been used to understand herbicide resistance's molecular basis. Moreover, cost-effective second-generation sequencing technologies (mainly RNA-seq) have become popular to identify the candidate genes involved in the NTSR. Genomics promises to go beyond the transcriptomic approaches and deliver additional insights into the genomic basis of resistance traits. For example, genome-based omics can provide insights into regulatory mechanisms involved in differential regulation of TSR and NTSR genes under herbicide stress. In addition to these, genomic resources are vital to answering questions related to copy number variation (CNVs). Copy number variation (CNV) of the herbicide target genes has repeatedly been shown to lie beneath herbicide resistance in several weed species (Salas et al., 2012; Wiersma et al., 2015; Lorentz et al., 2014). For a long time, it was believed that SNPs and insertion-deletion polymorphisms are the main types of DNA alterations, leading to the variation among the inter-species genomes. However, lately, the focus has also been given to the CNVs. CNVs are usually produced from unstable DNA modifications, increasing the number of copies of a particular DNA sequence. Besides triggering the overall production of the target proteins, if these short repeats occur within the non-coding portion of the gene, it might regulate gene expression levels (Żmieńko et al., 2014). Hence, it can be expected that CNVs will play substantial roles in plant evolution and adaptation under various stress conditions such as herbicide resistance. Current advances in genomics technologies enable researchers to develop several valuable methods to identify and report copy number variations within the genomes of non-model organisms. Hence, genome-based omics research will indisputably generate new perceptions of herbicide resistance mechanisms because the understanding of herbicide resistance in weeds is still in its infancy. However, with the beginning of a potential new era in weed science, genomic studies will offer many innovative strategies for managing weeds.

3.2.2 Understanding Epigenetic Mechanisms Involved in Resistance

Epigenetics modifications might be heritable traits that can be carried over multiple generations, thereby providing a selective advantage. Although the potential roles of epigenetic mechanisms in contributing to survival or fitness under various biotic and abiotic stresses are well documented, their contribution to herbicide resistance (especially in NTSR) in weeds is overlooked. Even though there are many well-documented reports on the various genetic changes that contribute to herbicide resistance in weeds, only a few have mentioned the consequences of the epigenetic mechanisms in developing herbicide resistance (Markus et al., 2018) (Table 3.1). Sublethal doses of herbicides induce various signalling pathways, activating general and stress-specific responses (Das et al., 2010). These responses control the expression and regulation of genes, which are directly or indirectly crucial for herbicide resistance. The signal generated by the herbicide may cause metabolite alteration that induces changes in the expression of modified enzymes, such as DNA methyltransferases, histone methyltransferases and histone deacetylases, that induce epigenetic changes. Thus, herbicide resistance may involve epigenetic mechanisms (Markus et al., 2018). The induced stable epigenetic effects

TABLE 3.1
Examples of Omics-Based Research Papers on Herbicide-Resistance

Omics Approach	Weed	Herbicide	Reference
Transcriptomics	*Beckmannia syzigachne*	Mesosulfuron-methyl	Wang, Chen et al., 2021
Transcriptomics	*Capsella bursa-pastoris*	Tribenuron-methyl	Zhang et al., 2021
Transcriptomics	*Solanum tuberosum*	Bentazone	Guo et al., 2021
Transcriptomics	*Conyza canadensis*	Halauxifen-methyl	McCauley et al., 2020
Transcriptomics	*Conyza bonariensis*	Glyphosate	Piasecki et al., 2019
Transcriptomics	*Polypogon fugax*	Clodinafop-propargyl	Zhou et al., 2017
Transcriptomics	*Lolium spp.*	Pyroxsulam	Duhoux et al., 2015
Transcriptomics	*Euphorbia esula*	Glyphosate	Doğramacı et al., 2014, 2016
Transcriptomics	*Lolium rigidum*	Diclofop-methyl	Gaines et al., 2014
Transcriptomics	*Soybean*	Glyphosate	Zhu et al., 2008
Epigenetics	*Zea mays*	Glyphosate	Tyczewska et al., 2021
Epigenetics	*Conyza canadensis*	Glyphosate	Margaritopoulou et al., 2018
Proteomics	*Alopecurus myosuroides*	Multiple herbicide resistant	Tétard-Jones et al., 2018
Proteomics	*Alopecurus aequalis*	Mesosulfuron-methyl	Zhao et al., 2019
Proteomics	*Echinochloa crus-galli L.*	Multiple herbicide resistant	Yang et al., 2017
Proteomics	*Chlamydomonas reinhardtii*	Multiple herbicide resistant	Nestler et al., 2012
Proteomics	*Zea mays*	Amiprophos methyl	Wang et al., 2011
Proteomics	*Sinapis arvensis L.*	Dicamba	Yajima et al., 2004
Metabolomics	*Lemna paucicostata*	Ascaulitoxin	Duke et al., 2011
Metabolomics	*Amaranthus palmeri*	Glyphosate	Maroli et al., 2015
Metabolomics	Multiple species	Saflufenacil	Grossmann et al., 2010

in weeds might become an inheritable trait after herbicide treatment, accelerating the evolution of herbicide resistance. In *Arabidopsis*, Ding et al. (2012) had shown that H3K4me3 acts as an obstinate epigenetic mark. They had shown that after multiple exposures to drought stress, the deposition of H3K4me3 in trainable genes is comparatively higher than that in nontrainable genes. Although some studies (Sani et al., 2013; Virlouvet et al., 2014) have shown that some epigenetic alterations might still exist even after recovery from stress and in the progeny, however, the mechanisms of inheriting these epigenetic changes as a stress memory mechanism still remained uncleared, to date (Sani et al., 2013; Virlouvet et al., 2014; Chang et al., 2020). Hence, untangling epigenetic codes related to the plant abiotic stress responses requires more attention in future studies.

Although advancements in the ultra-high-throughput techniques have improved our knowledge of the impacts of epigenetic changes on gene expression in model species, epigenetic studies related to herbicide resistance are still in their infancy. Given the increasing number of herbicide-resistant cases, epigenetics must be considered an essential regulatory mechanism. Preliminary studies in rice, wheat and *Arabidopsis* indicate that herbicides can trigger epigenetic responses. However, more experimental efforts are required to reveal the herbicide biochemistries to elucidate the epigenetic mechanisms of herbicide resistance. With the increase in epigenetic studies related to herbicide resistance, our understanding of weed adaptation to herbicides will open up new opportunities to understand the evolution of herbicide resistance in weeds.

3.3 RNA-BASED OMICS APPROACHES

3.3.1 Transcriptomics Abetting Understanding of Herbicide Resistance

Microarray technology has facilitated the possibility of examining thousands of genes simultaneously, thus providing a comprehensive view of gene activities under various environmental conditions. The major drawback of microarrays is that they are limited to providing relative abundance information about identified genes and have several issues related to the reproducibility of results. Moreover, microarrays require the knowledge of reference genome/transcriptome to design probes for the expression experiment. RNA-seq experiments have provided an excellent alternative to microarray-based strategies, as it involves sequencing the whole transcriptome of an organism. Therefore, reference genome/transcriptome sequence can be helpful to RNA-seq data analysis but is not necessary like microarray experiments (Vijay et al., 2013). Although challenges like developing methods and pipelines are common to RNA-seq experiments, its improved throughput processes and lower costs increased its potential to reveal the complexities within the plant transcriptome. Furthermore, to detect SNPs, splice variants, identification of novel genes and identification of transcript abundance upon DNA methylation-state modification remains an additional advantage for RNA-seq (Patterson et al., 2019). RNA-seq has a high detection range due to its digital nature and can even detect transcripts with shallow expression.

Although both TSR and NTSR mechanisms can be detected by RNA-seq, studying the entire transcriptome is particularly useful for NTSR mechanisms. This typically involves several genes and gene families (e.g., cytP450s, GSTs, GTs, ABC transporters, oxidases etc.). The first weed transcriptomes released were those of horseweed (*Erigeron canadensis L.*) (Peng et al., 2010) and waterhemp (*Amaranthus tuberculatus (Moq.) J. D. Sauer*) (Riggins et al., 2010). RNA-seq has been used to study responses to herbicides like glyphosate, ALS-inhibitors, ACCase-inhibitors, and auxinic inhibitors in several weed species like *Lolium rigidum* Gaudin (Gaines et al., 2014; Duhoux et al., 2015), *Alopecurus myosuroides* Huds. (Franco-Ortega et al., 2021), *Echinochloa crus-galli* (L.) P. Beauv. (Fang et al., 2019), *Beckmannia syzigachne* Steud. (Bai et al., 2020), *Bromus sterilis* L. (Sen et al., 2021 unpublished data) etc. Other examples involving transcriptomic studies can be found in Table 3.1.

High levels of heterozygosity and the presence of many complex gene families in the weeds have limited the assembling of a quality de novo reference transcriptome for weeds. Additionally, the polyploidy nature of weed further complicates the reference transcriptome assembly computation processes. RNA-seq detects all variations in a target gene expression and sequence. This becomes critical while studying herbicide resistance mechanisms. Using highly unrelated resistant (R) and susceptible (S) populations might lead to misleading results. Hence, even if some genes with significant differential expression are obtained, there might be a chance that they might be entirely unrelated to herbicide resistance. Henceforth, a high-quality reference transcriptome is needed to remove these false-positive genes.

3.3.2 Epitranscriptomic Regulation of Herbicide Resistance

The relation between the chemical modifications of DNA and proteins with the gene expression and stress responses has been well established in many cases. More recently, epi-transcriptome or RNA modifications have gained interest as an additional level of gene regulation that is deposited by "Writer proteins", recognized by "Reader proteins", and removed by "Eraser proteins". These modifications affecting the tRNA biogenesis may be of several types, depending on their position and the enzymes responsible. tRNAs are the most heavily modified types, and these modifications were found to be highly conserved among plants, animals and bacteria. RNA editing of adenosine to inosine, reduction of uridine to pseudouridine or dihydrouridine, methylation of the RNA bases and acetylation of the RNA bases are some of the few examples of tRNA modifications. Although the molecular functions of many of these tRNA modification events are still unclear, some of them have been well characterized (Hu et al., 2019; Vandivier & Gregory, 2018).

Target-site resistance mechanisms (TSR) involve a mutation within the target protein, which reduces or eliminates the binding of the herbicide. Therefore, it is relatively easy to study the molecular mechanisms of this type of resistance. However, non-target site-based resistance involves more than one mechanism, including overexpression of a particular gene or a group of gene families, changes to the translocation of herbicides within the plant and many others. Unravelling the molecular mechanisms behind non-target site-based resistance (NTSR) to herbicides is a major challenge for weed scientists. As of now, no study has been reported on epi-transcriptomic regulation of herbicide resistance. These RNA modifications can be used to investigate the mechanisms behind the non-target site-based resistance. It has been already established that epi-transcriptomic modifications are known to regulate both RNA quality as well as RNA quantity in eukaryotes. Hence, it can be assumed that these epi-transcriptome-mediated post-transcriptional modifications might profoundly affect weeds' development under different stresses and the responses of weeds to herbicide stresses. However, these assumptions need to be validated.

3.4 PROTEIN-BASED OMICS IN HERBICIDE RESISTANCE

The most challenging part of proteomics, or the large-scale study of proteins, is that proteomes vary from cell to cell and from time to time, which is entirely in contrast to the genomes, which remain almost constant within an organism. Abiotic stresses have profound impacts on plant proteomes. These effects include disparities in protein relative abundance, post-transcriptional modifications, post-translational modifications etc., which may shift the protein biological functions. Compared to earlier proteomics-based approaches such as two-dimensional electrophoresis (2-DE), modern techniques like isobaric tags for relative and absolute quantification (iTRAQ), along with liquid chromatography coupled with tandem mass spectrometry (LC-MS/MS), have emerged as a robust approach. iTRAQ proteomic technique enabled the researchers to explore a larger number of proteins at a large scale (almost four-fold more than conventional gel-based methods), thus providing specific information on protein isoforms besides detecting the low-abundance proteins. Although proteomics-based studies related to herbicide resistance have been reported in some weed species such as *Echinochloa crus-galli* L. (Yang et al., 2017), *Avena fatua* L. (Burns et al., 2017), *A. myosuroides* (Tétard-Jones et al., 2018) and *L. multiflorum* (Barroso et al., 2018), proteomic research into herbicide stress responses in arable weeds is still limited (Table 3.1). The exposure of herbicide-resistant weeds to the herbicide stress induces an adequate stress response, which might reorganize the plant's proteome. Hence, comparative proteomic studies between susceptible and herbicide-resistant weeds can substantially contribute to identifying novel potential protein markers of herbicide stress tolerance. These protein markers will accelerate further research and applications in weed management. However, before further applications, validation with protein quantification methods such as ELISA or immunoblots must be conducted.

Besides proteomics, epi-proteomics and interactomics can also be considered emerging prospective research areas for herbicide resistance. Epi-proteomics includes changes in proteins resulting from epigenetic modifications, such as histone acetylation and deacetylation. Various modifications in the histone proteins, like methylation, acetylation, phosphorylation, etc., are critical in plants' survival under abiotic stresses (Zheng et al., 2016). Similarly, the physical interactions between different proteins can be imperative for gene function. Protein–protein interactions have always played important roles in orchestrating biological processes such as signal transduction, stress responses and plant defences. Protein–protein interaction could be significant at the molecular level in recruiting various transcriptional co-factors and enzymes responsible for post-transcriptional modifications and transporter activation (Zhang et al., 2010). Uncovering the epi-proteomic and interactomics responses to herbicide stress will contribute to our insights into how weeds adapt to herbicide stress; however, such research is currently underdeveloped in weed science.

3.5 METABOLITES-BASED OMICS TO DECODE HERBICIDE RESISTANCE MECHANISM

Among the latest 'omics' sciences, metabolomics has earned a high reputation among plant biologists (Kumar et al., 2017). Metabolomics is a powerful tool that can detect minor differences between the levels of thousands of molecules between systems exposed to various treatments. Moreover, considering the developments in informatics and analytical technologies, metabolomics has the potential to become an integral tool for biomarker discovery. Hence, this 'omics' science can be employed to study weeds' response to herbicide applications in detail. Although, in the past, metabolomics has been used widely to study the MOAs of synthetic and natural herbicidal compounds, its application to characterize weed physiology in response to herbicide applications has been minimal. Some examples of the use of metabolomics can be found in *Amaranthus* sp. (Maroli et al., 2015, 2016), *Ipomoea* sp. (Maroli, Gaines et al., 2018) etc., and in Table 3.1. In the eukaryotes, small molecules, the polymer-based composition of various organelles and the integration of various metabolic reactions play a crucial role in generating and maintaining the identity of a cell. Despite the availability of genes, transcripts and proteins, the information had failed to explain many physiological responses. Therefore, the gap between gene expression and the phenotypic outcome is self-evident for biological processes. Metabolomics can comprehensively contribute to molecular phenotyping, reducing the existing gap (Maroli, Nandula et al., 2018), especially in agrobiology.

The response of weeds to various agrochemicals can be studied in detail by the metabolomic approach, which enables monitoring of metabolic changes in response to stresses such as herbicides. Moreover, toxicological and ecotoxicological risks can also be assessed for new and modified existing bioactive compounds. Hence, it can be expected that advanced metabolomics studies, combined with other omics-based and biochemical studies, will contribute to the various areas of weed science, including herbicide-resistant mechanism elucidation and characterization of weed response to the herbicide.

3.6 GENE-DRIVING STRATEGIES TO TOPPLE HERBICIDE RESISTANCE

Genome editing can provide heritable, targeted-modified and transgenic-free crops. Gene-editing techniques have been already applied effectively in major crops to increase their yield (Gupta et al., 2019; Razzaq et al., 2021). Besides being an eco-friendly and practicable option, this has become an efficient option to meet the increasing global food demand. Additionally, unlike transgenic approaches, plant genome engineering is exempted from ethical issues in many countries (although some countries consider gene editing transgenic) (Razzaq et al., 2021; Hussain et al., 2021). Recently, many research groups have used gene-editing techniques like CRISPER-Cas9, Transcription activator-like effector nucleases (TALEN), Cytidine base editor (CBE), prime editing etc. successfully to develop herbicide-resistant plants (Sun et al., 2016; Wang, Xu et al., 2021; Hussain et al., 2021). Various research studies conducted to develop herbicide-resistant plants through genome editing can be found in Hussain et al. (2021).

To date, site-specific genome editing has remained a significant challenge in many eukaryotic species, including plants. However, despite some initial success with gene-editing technologies, the efficacy of template DNA delivery in non-model species like weeds is quite challenging. The lack of efficiency restricts the ability to create creating specific mutations in the DNA sequence. Nevertheless, these limitations can be delimited with more research in this area.

3.7 CONCLUDING REMARKS AND FUTURE DIRECTIONS

In the last two decades, omics-based technologies such as genomics, metabolomics, transcriptomics and proteomics, powered by advanced statistical and computational methodologies and nucleic acid and peptide sequencing platforms, have been used to elucidate molecular mechanisms of plant responses to biotic and abiotic stresses. Moreover, recently, omics-based technologies have successfully devised new strategies to determine new modes of action. However, to get a complete system-level understanding, data integrations from various omics-based technologies can be a valuable strategy. To facilitate future omics-based studies, significant efforts should be offered to publish genomes of essential weed species. High throughput techniques related to gene expression, protein and metabolic profiling allow the combination of data to develop a holistic understanding of the model and non-model plant species. Although basic plant biology has already integrated various omics technologies, their integration in weed biology is still in infancy. The future integration of phenomic data with other multi-omics data will allow weed biologists to address various herbicide resistance mechanisms more efficiently. Gene-drive based strategies (such as CRISPR) can also be considered for sustainable weed management. Relying on only omics-based technologies might not be enough in the future, as weeds might employ more precise and varying mechanisms to adapt to herbicide stress. Eukaryotic cells have highly specialized functional units that are reaction vessels composed of various integrated systems. The journey from gene to protein and metabolites is far more complex, integrated and multifactorial.

Moreover, varying levels of stability between the RNAs and the proteins make the comparison between gene expression levels and protein abundance complicated. With the improvement of the techniques related to gene expression profiling and proteomic profiling, their data can be combined to get a holistic understanding of the model and non-model plant species. Even though basic plant biology has already integrated various omics technologies, their integration in weed biology is still at its infinity. The future integration of phenomic data with other multi-omics data will allow weed biologists to address various herbicide resistance mechanisms in a more efficient/effective manner. Hence, combinatorial approaches will be needed, and gene-editing techniques can be applied.

In conclusion, despite substantial challenges in applying omics to weeds, omics-based technologies can yield promising results in explaining fundamental weed biology concepts. The increase in the data acquisition and development of new statistical approaches can overcome existing challenges with big data handling. Furthermore, with the help of these omics platforms, the molecular mechanisms involved in response to biotic and abiotic stresses and crop–weed competition can also be elucidated, leading to the plausible formulation of long-term, sustainable weed management solutions.

ACKNOWLEDGEMENT

MKS, KH, PK, JS acknowledge the Ministry of Agriculture of the Czech Republic, Project No. QK1820081, the Department of Agroecology and Crop Production, Faculty of Agrobiology, Food and Natural Resources, Czech University of Life Sciences Prague, for infrastructural and financial support. AR would like to acknowledge "EVA4.0", No. CZ.02.1.01/0.0/0.0/16_019/0000803 financed by "OP RDE" and "EXTEMIT – K", No. CZ.02.1.01/0.0/0.0/15_003/0000433 financed by OP RDE and Excellent Team Grant (2023-2024), Faculty of Forestry and Wood Sciences, Czech University of Life Sciences, Prague, for the financial support during the preparation of the book chapter.

REFERENCES

Abou-Nader, M., & Benedik, M. J. (2010). Rapid generation of random mutant libraries. *Bioengineered Bugs*, *1*(5), 337–340. https://doi.org/10.4161/bbug.1.5.12942

Baerson, S. R., Rodriguez, D. J., Biest, N. A., Tran, M., You, J., Kreuger, R. W., Dill, G. M., Pratley, J. E., & Gruys, K. J. (2002). Investigating the mechanism of glyphosate resistance in rigid ryegrass (*Lolium ridigum*). *Weed Science*, *50*(6), 721–730. https://doi.org/10.1614/0043-1745(2002)050[0721:ITMOGR]2.0.CO;2

Bai, S., Zhao, Y., Zhou, Y., Wang, M., Li, Y., Luo, X., & Li, L. (2020). Identification and expression of main genes involved in non-target site resistance mechanisms to fenoxaprop-p-ethyl in *Beckmannia syzigachne*. *Pest Management Science*, *76*(8), 2619–2626. https://doi.org/10.1002/ps.5800

Barroso, A. A., de S Costa, M. G., Neto, N. J., Dos Santos, J. I., Balbuena, T. S., Carbonari, C. A., & Alves, P. L. (2018). Protein identification before and after glyphosate exposure in *Lolium multiflorum* genotypes. *Pest Management Science*, *74*(5), 1125–1133. https://doi.org/10.1111/tpj.13892

Baucom, R. S. (2019). Evolutionary and ecological insights from herbicide-resistant weeds: What have we learned about plant adaptation, and what is left to uncover? *The New Phytologist*, *223*(1), 68–82. https://doi.org/10.1111/nph.15723

Burns, E. E., Keith, B. K., Refai, M. Y., Bothner, B., & Dyer, W. E. (2017). Proteomic and biochemical assays of glutathione-related proteins in susceptible and multiple herbicide resistant *Avena fatua* L. *Pesticide Biochemistry and Physiology*, *140*, 69–78. https://doi.org/10.1016/j.pestbp.2017.06.007

Chang, Y. N., Zhu, C., Jiang, J., Zhang, H., Zhu, J. K., & Duan, C. G. (2020). Epigenetic regulation in plant abiotic stress responses. *Journal of Integrative Plant Biology*, *62*(5), 563–580. https://doi.org/10.1111/jipb.12901

Christopher, J. T., Powles, S. B., Liljegren, D. R., & Holtum, J. A. M. (1991). Cross-resistance to herbicides in annual ryegrass (*Lolium rigidum*) 1: II. Chlorsulfuron resistance involves a wheat-like detoxification system. *Plant Physiology*, *95*(4), 1036–1043. https://doi.org/10.1104/pp.95.4.1036

Das, M., Reichman, J. R., Haberer, G., Welzl, G., Aceituno, F. F., Mader, M. T., Watrud, L. S., Pfleeger, T. G., Gutiérrez, R. A., Schäffner, A. R., & Olszyk, D. M. (2010). A composite transcriptional signature differentiates responses towards closely related herbicides in *Arabidopsis thaliana* and *Brassica napus*. *Plant Molecular Biology*, *72*(4–5), 545–556. https://doi.org/10.1007/s11103-009-9590-y

Délye, C., Gardin, J. A. C., Boucansaud, K., Chauvel, B., & Petit, C. (2011). Non-target-site-based resistance should be the centre of attention for herbicide resistance research: *Alopecurus myosuroides* as an illustration. *Weed Research*, *51*(5), 433–437. https://doi.org/10.1111/j.1365-3180.2011.00864.x

Délye, C., Jasieniuk, M., & Le Corre, V. (2013). Deciphering the evolution of herbicide resistance in weeds. *Trends in Genetics*, *29*(11), 649–658. https://doi.org/10.1016/j.tig.2013.06.001

Ding, Y., Fromm, M., & Avramova, Z. (2012). Multiple exposures to drought "train" transcriptional responses in *Arabidopsis*. *Nature Communications*, *3*(1), 1–9. https://doi.org/10.1038/ncomms1732

Doğramacı, M., Anderson, J. V., Chao, W. S., & Foley, M. E. (2014). Foliar application of glyphosate affects molecular mechanisms in underground adventitious buds of leafy spurge (*Euphorbia esula*) and alters their vegetative growth patterns. *Weed Science*, *62*(2), 217–229. https://doi.org/10.1614/WS-D-13-00156.1

Doğramacı, M., Gramig, G. G., Anderson, J. V., Chao, W. S., & Foley, M. E. (2016). Field application of glyphosate induces molecular changes affecting vegetative growth processes in leafy spurge (*Euphorbia esula*). *Weed Science*, *64*(1), 87–100. https://doi.org/10.1614/WS-D-15-00108.1

Duhoux, A., Carrère, S., Gouzy, J., Bonin, L., & Délye, C. (2015). RNA-Seq analysis of rye-grass transcriptomic response to an herbicide inhibiting acetolactate-synthase identifies transcripts linked to non-target-site-based resistance. *Plant Molecular Biology*, *87*(4–5), 473–487. https://doi.org/10.1007/s11103-015-0292-3

Duke, S. O., Evidente, A., Fiore, M., Rimando, A. M., Dayan, F. E., Vurro, M., Christiansen, N., Looser, R., Hutzler, J., & Grossmann, K. (2011). Effects of the aglycone of ascaulitoxin on amino acid metabolism in *Lemna paucicostata*. *Pesticide Biochemistry and Physiology*, *100*(1), 41–50. https://doi.org/10.1016/j.pestbp.2011.02.002

Fang, J., Zhang, Y., Liu, T., Yan, B., Li, J., & Dong, L. (2019). Target-site and metabolic resistance mechanisms to penoxsulam in barnyardgrass (*Echinochloa crus-galli* (L.) P. Beauv). *Journal of Agricultural and Food Chemistry*, *67*(29), 8085–8095. https://doi.org/10.1021/acs.jafc.9b01641

Franco-Ortega, S., Goldberg-Cavalleri, A., Walker, A., Brazier-Hicks, M., Onkokesung, N., & Edwards, R. (2021). Non-target site herbicide resistance is conferred by two distinct mechanisms in black-grass (*Alopecurus myosuroides*). *Frontiers in Plant Science*, *12*, 194. https://doi.org/10.3389/fpls.2021.636652

Gaines, T. A., Duke, S. O., Morran, S., Rigon, C. A. G., Tranel, P. J., Küpper, A., & Dayan, F. E. (2020). Mechanisms of evolved herbicide resistance. *Journal of Biological Chemistry*. https://doi.org/10.1074/jbc.REV120.013572

Gaines, T. A., Lorentz, L., Figge, A., Herrmann, J., Maiwald, F., Ott, M.-C., Han, H., Busi, R., Yu, Q., Powles, S. B., & Beffa, R. (2014). RNA-Seq transcriptome analysis to identify genes involved in metabolism-based diclofop resistance in *Lolium rigidum*. *The Plant Journal*, *78*(5), 865–876. https://doi.org/10.1111/tpj.12514

Ghanizadeh, H., & Harrington, K. C. (2017). Non-target site mechanisms of resistance to herbicides. *Critical Reviews in Plant Sciences*, *36*(1), 24–34. https://doi.org/10.1080/07352689.2017.1316134

Gharde, Y., Singh, P. K., Dubey, R. P., & Gupta, P. K. (2018). Assessment of yield and economic losses in agriculture due to weeds in India. *Crop Protection*, *107*, 12–18. https://doi.org/10.1016/j.cropro.2018.01.007

Grossmann, K., Niggeweg, R., Christiansen, N., Looser, R., & Ehrhardt, T. (2010). The herbicide saflufenacil (Kixor™) is a new inhibitor of protoporphyrinogen IX oxidase activity. *Weed Science*, *58*(1), 1–9.

Guo, J., Song, X., Sun, S., Shao, B., Tao, B., & Zhang, L. (2021). RNA-Seq transcriptome analysis of potato with differential tolerance to bentazone herbicide. *Agronomy*, *11*(5), 897. https://doi.org/10.3390/agronomy11050897

Gupta, D., Bhattacharjee, O., Mandal, D., Sen, M. K., Dey, D., Dasgupta, A., Kazi, T. A., Gupta, R., Sinharoy, S., Acharya, K., Chattopadhyay, D., Ravichandiran, V., Roy, S., & Ghosh, D. (2019). CRISPR-Cas9 system: A new-fangled dawn in gene editing. *Life Sciences*, *232*, 116636. https://doi.org/10.1016/j.lfs.2019.116636

Hall, L. M., Holtum, J. A., & Powles, S. B. (2018). Mechanisms responsible for cross resistance and multiple resistance. In *Herbicide resistance in plants* (pp. 243–262). CRC Press.

Hall, L. M., Moss, S. R., & Powles, S. B. (1995). Mechanism of resistance to chlorotoluron in two biotypes of the grass weed *Alopecurus myosuroides*. *Pesticide Biochemistry and Physiology*, *53*(3), 180–192. https://doi.org/10.1006/pest.1995.1066

Haughn, G. W., Smith, J., Mazur, B., & Somerville, C. (1988). Transformation with a mutant *Arabidopsis* acetolactate synthase gene renders tobacco resistant to sulfonylurea herbicides. *Molecular and General Genetics MGG*, *211*(2), 266–271. https://doi.org/10.1007/BF00330603

Horvath, D. (2010). Genomics for weed science. *Current Genomics*, *11*(1), 47–51. https://doi.org/10.2174/138920210790217972

Hu, J., Manduzio, S., & Kang, H. (2019). Epitranscriptomic RNA methylation in plant development and abiotic stress responses. *Frontiers in Plant Science*, *10*, 500. https://doi.org/10.3389/fpls.2019.00500

Hussain, A., Ding, X., Alariqi, M., Manghwar, H., Hui, F., Li, Y., Cheng, J., Wu, C., Cao, J., & Jin, S. (2021). Herbicide resistance: Another hot agronomic trait for plant genome editing. *Plants*, *10*(4), 621. https://doi.org/10.3390/plants10040621

Košnarová, P., Hamouz, P., Hamouzová, K., Linn, A., Sen, M. K., Mikulka, J., Šuk, J., & Soukup, J. (2021). *Apera spica-venti* in the Czech Republic develops resistance to three herbicide modes of action. *Weed Research*. https://doi.org/10.1111/wre.12500

Kumar, R., Bohra, A., Pandey, A. K., Pandey, M. K., & Kumar, A. (2017). Metabolomics for plant improvement: Status and prospects. *Frontiers in Plant Science*, *8*, 1302. https://doi.org/10.3389/fpls.2017.01302

Lorentz, L., Gaines, T. A., Nissen, S. J., Westra, P., Strek, H. J., Dehne, H. W., Ruiz-Santaella, J. P., & Beffa, R. (2014). Characterization of glyphosate resistance in *Amaranthus tuberculatus* populations. *Journal of Agricultural and Food Chemistry*, *62*(32), 8134–8142. https://doi.org/10.1021/jf501040x

Lorraine-Colwill, D. F., Powles, S. B., Hawkes, T. R., Hollinshead, P., Warner, S. A. J., & Preston, C. (2002). Investigations into the mechanism of glyphosate resistance in *Lolium rigidum*. *Pesticide Biochemistry and Physiology*, *74*(2), 62–72. https://doi.org/10.1016/S0048-3575(03)00007-5

Margaritopoulou, T., Tani, E., Chachalis, D., & Travlos, I. (2018). Involvement of epigenetic mechanisms in herbicide resistance: The case of *Conyza canadensis*. *Agriculture*, *8*(1), 17. https://doi.org/10.3390/agriculture8010017

Markus, C., Pecinka, A., Karan, R., Barney, J. N., & Merotto Jr, A. (2018). Epigenetic regulation – contribution to herbicide resistance in weeds? *Pest Management Science*, *74*(2), 275–281. https://doi.org/10.1002/ps.4727

Maroli, A. S., Gaines, T. A., Foley, M. E., Duke, S. O., Doğramacı, M., Anderson, J. V., Horvath, D. P., Chao, W. S., & Tharayil, N. (2018). Omics in weed science: A perspective from genomics, transcriptomics, and metabolomics approaches. *Weed Science*, *66*(6), 681–695. https://doi.org/10.1017/wsc.2018.33

Maroli, A. S., Nandula, V. K., Dayan, F. E., Duke, S. O., Gerard, P., & Tharayil, N. (2015). Metabolic profiling and enzyme analyses indicate a potential role of antioxidant systems in complementing glyphosate resistance in an *Amaranthus palmeri* biotype. *Journal of Agricultural and Food Chemistry*, *63*(41), 9199–9209. https://doi.org/10.1021/acs.jafc.5b04223

Maroli, A. S., Nandula, V. K., Duke, S. O., Gerard, P., & Tharayil, N. (2018). Comparative metabolomic analyses of *Ipomoea lacunosa* biotypes with contrasting glyphosate tolerance captures herbicide-induced differential perturbations in cellular physiology. *Journal of Agricultural and Food Chemistry*, *66*(8), 2027–2039. https://doi.org/10.1021/acs.jafc.7b04722

Maroli, A. S., Nandula, V. K., Duke, S., & Tharayil, N. (2016). Stable isotope resolved metabolomics reveals the role of anabolic and catabolic processes in glyphosate-induced amino acid accumulation in *Amaranthus palmeri* biotypes. *Journal of Agricultural and Food Chemistry*, *64*(37), 7040–7048. https://doi.org/10.1021/acs.jafc.6b02196

McCauley, C. L., McAdam, S. A., Bhide, K., Thimmapuram, J., Banks, J. A., & Young, B. G. (2020). Transcriptomics in *Erigeron canadensis* reveals rapid photosynthetic and hormonal responses to auxin herbicide application. *Journal of Experimental Botany*, *71*(12), 3701–3709. https://doi.org/10.1093/jxb/eraa124

Montgomery, J. S., Giacomini, D., Waithaka, B., Lanz, C., Murphy, B. P., Campe, R., Lerchl, J., Landes, A., Gatzmann, F., Janssen, A., Antonise, R., Patterson, E., Weigel, D., & Tranel, P. J. (2020). Draft Genomes of *Amaranthus tuberculatus, Amaranthus hybridus, and Amaranthus palmeri*. *Genome Biology and Evolution*, *12*(11), 1988–1993. https://doi.org/10.1093/gbe/evaa177

Moss, S. R. (2002). Herbicide-resistant weeds. In *Weed management handbook* (pp. 225–252). John Wiley & Sons, Ltd. https://doi.org/10.1002/9780470751039.ch11

Nestler, H., Groh, K. J., Schönenberger, R., Eggen, R. I., & Suter, M. J. F. (2012). Linking proteome responses with physiological and biochemical effects in herbicide-exposed Chlamydomonas reinhardtii. *Journal of Proteomics*, *75*(17), 5370–5385.

Owen, M. J., Walsh, M. J., Llewellyn, R. S., & Powles, S. B. (2007). Widespread occurrence of multiple herbicide resistance in Western Australian annual ryegrass (*Lolium rigidum*) populations. *Australian Journal of Agricultural Research*, *58*(7), 711–718. https://doi.org/10.1071/AR06283

Patterson, E. L., Saski, C., Küpper, A., Beffa, R., & Gaines, T. A. (2019). Omics potential in herbicide-resistant weed management. *Plants*, *8*(12), 607. https://doi.org/10.3390/plants8120607

Patzoldt, W. L., Hager, A. G., McCormick, J. S., & Tranel, P. J. (2006). A codon deletion confers resistance to herbicides inhibiting protoporphyrinogen oxidase. *Proceedings of the National Academy of Sciences*, *103*(33), 12329–12334. https://doi.org/10.1073/pnas.0603137103

Peng, Y., Abercrombie, L. L., Yuan, J. S., Riggins, C. W., Sammons, R. D., Tranel, P. J., & Stewart, C. N. (2010). Characterization of the horseweed (Conyza canadensis) transcriptome using GS-FLX 454 pyrosequencing and its application for expression analysis of candidate non-target herbicide resistance genes. *Pest management Science*, *66*(10), 1053–1062. https://doi.org/10.1002/ps.2004

Piasecki, C., Yang, Y., Benemann, D. P., Kremer, F. S., Galli, V., Millwood, R. J., Cechin, J., Agostinetto, D., Maia, L. C., Vargas, L., & Stewart, C. N. (2019). Transcriptomic analysis identifies new non-target site glyphosate-resistance genes in Conyza bonariensis. *Plants*, *8*(6), 157. https://doi.org/10.3390/plants8060157

Ravet, K., Patterson, E. L., Krähmer, H., Hamouzová, K., Fan, L., Jasieniuk, M., Lawton-Rauh, A., Malone, J. M., McElroy, J. S., Merotto, A., Westra, P., & Gaines, T. A. (2018). The power and potential of genomics in weed biology and management. *Pest Management Science*, *74*(10), 2216–2225. https://doi.org/10.1002/ps.5048

Razzaq, M. K., Aleem, M., Mansoor, S., Khan, M. A., Rauf, S., Iqbal, S., & Siddique, K. H. M. (2021). Omics and CRISPR-Cas9 approaches for molecular insight, functional gene analysis, and stress tolerance development in crops. *International Journal of Molecular Sciences*, *22*(3), 1292. https://doi.org/10.3390/ijms22031292

Reade, J. P. H., & Cobb, A. H. (2002). Herbicides: Modes of action and metabolism. In *Weed management handbook* (pp. 134–170). John Wiley & Sons, Ltd. https://doi.org/10.1002/9780470751039.ch8

Rey-Caballero, J., Menéndez, J., Osuna, M. D., Salas, M., & Torra, J. (2017). Target-site and non-target-site resistance mechanisms to ALS inhibiting herbicides in Papaver rhoeas. *Pesticide Biochemistry and Physiology*, *138*, 57–65. https://doi.org/10.1016/j.pestbp.2017.03.001

Riggins, C. W., Peng, Y., Stewart Jr, C. N., & Tranel, P. J. (2010). Characterization of de novo transcriptome for waterhemp (*Amaranthus tuberculatus*) using GS-FLX 454 pyrosequencing and its application for studies of herbicide target-site genes. *Pest Management Science*, *66*(10), 1042–1052. https://doi.org/10.1002/ps.2006

Salas, R. A., Dayan, F. E., Pan, Z., Watson, S. B., Dickson, J. W., Scott, R. C., & Burgos, N. R. (2012). EPSPS gene amplification in glyphosate-resistant Italian ryegrass (Lolium perenne ssp. multiflorum) from Arkansas. *Pest Management Science*, *68*(9), 1223–1230. https://doi.org/10.1002/ps.3342

Sani, E., Herzyk, P., Perrella, G., Colot, V., & Amtmann, A. (2013). Hyperosmotic priming of *Arabidopsis* seedlings establishes a long-term somatic memory accompanied by specific changes of the epigenome. *Genome Biology*, *14*(6), 1–24. https://doi.org/10.1186/gb-2013-14-6-r59

Sharma, A., Kumar, V., Shahzad, B., Tanveer, M., Sidhu, G. P. S., Handa, N., Kohli, S. K., Yadav, P., Bali, A. S., Parihar, R. D., Dar, O. I., Singh, K., Jasrotia, S., Bakshi, P., Ramakrishnan, M., Kumar, S., Bhardwaj, R., & Thukral, A. K. (2019). Worldwide pesticide usage and its impacts on ecosystem. *SN Applied Sciences*, *1*(11), 1446. https://doi.org/10.1007/s42452-019-1485-1

Sen, M. K., Hamouzová, K., Mikulka, J., Bharati, R., Košnarová, P., Hamouz, P., Roy, A., & Soukup, J. (2021). Enhanced metabolism and target gene overexpression confer resistance against acetolactate synthase-inhibiting herbicides in Bromus sterilis. *Pest Management Science*, *77*(4), 2122–2128. https://doi.org/10.1002/ps.6241

Sun, Y., Zhang, X., Wu, C., He, Y., Ma, Y., Hou, H., Guo, X., Du, W., Zhao, Y., & Xia, L. (2016). Engineering herbicide-resistant rice plants through CRISPR/Cas9-mediated homologous recombination of acetolactate synthase. *Molecular Plant*, *9*(4), 628–631. https://doi.org/10.1016/j.molp.2016.01.001

Tétard-Jones, C., Sabbadin, F., Moss, S., Hull, R., Neve, P., & Edwards, R. (2018). Changes in the proteome of the problem weed blackgrass correlating with multiple-herbicide resistance. *The Plant Journal*, *94*(4), 709–720. https://doi.org/10.1111/tpj.13892

Tyczewska, A., Gracz-Bernaciak, J., Szymkowiak, J., & Twardowski, T. (2021). Herbicide stress-induced DNA methylation changes in two Zea mays inbred lines differing in Roundup® resistance. *Journal of Applied Genetics*, *62*(2), 235–248. https://doi.org/10.1007/s13353-021-00609-4

Vandivier, L. E., & Gregory, B. D. (2018). New insights into the plant epitranscriptome. *Journal of Experimental Botany*, *69*(20), 4659–4665. https://doi.org/10.1093/jxb/ery262

Vazquez-Garcia, J. G., Alcantara-de la Cruz, R., Rojano-Delgado, A. M., Palma-Bautista, C., de Portugal Vasconcelos, J. M., & De Prado, R. (2021). Multiple herbicide resistance evolution: The case of Eleusine indica in Brazil. *Journal of Agricultural and Food Chemistry*, *69*(4), 1197–1205. https://doi.org/10.1021/acs.jafc.0c03999

Vijay, N., Poelstra, J. W., Künstner, A., & Wolf, J. B. (2013). Challenges and strategies in transcriptome assembly and differential gene expression quantification: A comprehensive in silico assessment of RNA-seq experiments. *Molecular Ecology*, *22*(3), 620–634. https://doi.org/10.1111/mec.12014

Vila-Aiub, M. M., Gundel, P. E., Yu, Q., & Powles, S. B. (2013). Glyphosate resistance in Sorghum halepense and Lolium rigidum is reduced at suboptimal growing temperatures. *Pest Management Science*, *69*(2), 228–232. https://doi.org/10.1002/ps.3464

Virlouvet, L., Ding, Y., Fujii, H., Avramova, Z., & Fromm, M. (2014). ABA signaling is necessary but not sufficient for RD 29 B transcriptional memory during successive dehydration stresses in A rabidopsis thaliana. *The Plant Journal*, *79*(1), 150–161. https://doi.org/10.1111/tpj.12548

Wang, F., Xu, Y., Li, W., Chen, Z., Wang, J., Fan, F., Tao, Y., Jiang, Y., Zhu, Q.-H., & Yang, J. (2021). Creating a novel herbicide-tolerance OsALS allele using CRISPR/Cas9-mediated gene editing. *The Crop Journal*, *9*(2), 305–312. https://doi.org/10.1016/j.cj.2020.06.001

Wang, J., Chen, J., Li, X., & Cui, H. (2021). RNA-Seq transcriptome analysis to identify candidate genes involved in non-target site-based mesosulfuron-methyl resistance in *Beckmannia syzigachne*. *Pesticide Biochemistry and Physiology*, *171*, 104738. https://doi.org/10.1016/j.pestbp.2020.104738

Wang, Z., Li, Q., Zhao, J., & Peng, Y. (2011). Investigation of the effect of herbicide amiprophos methyl on spindle formation and proteome change in maize by immunofluorescence and proteomic technique. *Cytologia*, *76*(3), 249–259. https://doi.org/10.1508/cytologia.76.249

Wiersma, A. T., Gaines, T. A., Preston, C., Hamilton, J. P., Giacomini, D., Robin Buell, C., Leach, J. E., & Westra, P. (2015). Gene amplification of 5-enol-pyruvylshikimate-3-phosphate synthase in glyphosate-resistant Kochia scoparia. *Planta*, *241*(2), 463–474. https://doi.org/10.1007/s00425-014-2197-9

Yajima, W., Hall, J. C., & Kav, N. N. (2004). Proteome-level differences between auxinic-herbicide-susceptible and-resistant wild mustard (Sinapis arvensis L.). *Journal of Agricultural and Food Chemistry*, *52*(16), 5063–5070. https://doi.org/10.1021/jf0495284

Yang, X., Zhang, Z., Gu, T., Dong, M., Peng, Q., Bai, L., & Li, Y. (2017). Quantitative proteomics reveals ecological fitness cost of multi-herbicide resistant barnyardgrass (*Echinochloa crus-galli* L.). *Journal of Proteomics*, *150*, 160–169. https://doi.org/10.1016/j.jprot.2016.09.009

You, C., & Zhang, Y. H. P. (2012). Easy preparation of a large-size random gene mutagenesis library in Escherichia coli. *Analytical Biochemistry*, *428*(1), 7–12. https://doi.org/10.1016/j.ab.2012.05.022

Yu, Q., Cairns, A., & Powles, S. B. (2004). Paraquat resistance in a population of Lolium rigidum. *Functional Plant Biology*, *31*(3), 247–254. https://doi.org/10.1071/fp03234

Yu, Q., Nelson, J. K., Zheng, M. Q., Jackson, M., & Powles, S. B. (2007). Molecular characterization of resistance to ALS-inhibiting herbicides in Hordeum leporinum biotypes. *Pest Management Science*, *63*(9), 918–927. https://doi.org/10.1002/ps.1429

Yuan, J. S., Tranel, P. J., & Stewart, C. N. (2007). Non-target-site herbicide resistance: A family business. *Trends in Plant Science*, *12*(1), 6–13. https://doi.org/10.1016/j.tplants.2006.11.001

Zhang, Y., Gao, P., & Yuan, J. S. (2010). Plant protein-protein interaction network and interactome. *Current Genomics*, *11*(1), 40–46. https://doi.org/10.2174/138920210790218016

Zhang, X., Wang, H., Bei, F., Wu, C., Zhang, L., Jia, S., Wang, J., & Liu, W. (2021). Investigating the mechanism of metabolic resistance to tribenuron-methyl in *capsella bursa-pastoris* (L.) Medik. by Full-length transcriptome assembly combined with RNA-Seq. *Journal of Agricultural and Food Chemistry*, *69*(12), 3692–3701. https://doi.org/10.1021/acs.jafc.0c07512

Zhao, N., Yan, Y., Luo, Y., Zou, N., Liu, W., & Wang, J. (2019). Unravelling mesosulfuron-methyl phytotoxicity and metabolism-based herbicide resistance in Alopecurus aequalis: Insight into regulatory mechanisms using proteomics. *Science of the Total Environment*, *670*, 486–497. https://doi.org/10.1016/j.scitotenv.2019.03.089

Zheng, Y., Huang, X., & Kelleher, N. L. (2016). Epiproteomics: Quantitative analysis of histone marks and codes by mass spectrometry. *Current Opinion in Chemical Biology*, *33*, 142–150. https://doi.org/10.1016/j.cbpa.2016.06.007

Zhou, F., Zhang, Y., Tang, W., Wang, M., & Gao, T. (2017). Transcriptomics analysis of the flowering regulatory genes involved in the herbicide resistance of Asia minor bluegrass (*Polypogon fugax*). *BMC Genomics*, *18*(1), 1–13. https://doi.org/10.1186/s12864-017-4324-z

Zhu, J., Patzoldt, W. L., Shealy, R. T., Vodkin, L. O., Clough, S. J., & Tranel, P. J. (2008). Transcriptome response to glyphosate in sensitive and resistant soybean. *Journal of Agricultural and Food Chemistry*, *56*(15), 6355–6363. https://doi.org/10.1021/jf801254e

Żmieńko, A., Samelak, A., Kozłowski, P., & Figlerowicz, M. (2014). Copy number polymorphism in plant genomes. *Theoretical and Applied Genetics*, *127*(1), 1–18. https://doi.org/10.1007/s00122-013-2177-7

4 Impact of Genomics in Improvement of Rice Bean Production and Productivity

An Orphan Crop

Bhupender Singh, Dibyalochan Mohanty, and Atul Kumar Upadhyay

4.1 INTRODUCTION

Grains and legumes are the central pillars for global distribution of nutritionally enriched food. They hold a special position in the diet of the majority of the population residing in developing countries. Legumes are highly nutritious in terms of proteins, vitamins, starch, minerals and oils (Lolas & Markakis, 1975). *Vigna umbellate*, commonly known as rice bean, is a grain legume is harvested worldwide, especially in mountain ranges, and poor farmers rely on this type of crop for their sustainability. Despite this, the economic importance of this particular crop is little known. Although the plant is highly nutritious and producible, its role in advancing the lives of the deprived is still not fully understood (Katoch, 2013). Rice bean is easily harvestable, with a short maturation period, and offers a high yield. Apart from being a good nutritional source for both humans and animals, it also helps to advance potential soil fertility (Echo Development Notes, issue 83, 2004).

4.2 BASIC BIOLOGY, CULTIVATION PRACTICES AND MAJOR LIMITATIONS IN CULTIVATION

The rice bean plant looks like a small vine, with hairy-looking yellow flowers. Following the maturation period, the flowers become consumable beans. The color of the seeds changes, primarily to something between yellow and red, but can also be black, brown, violet or ivory. Rice bean is efficient and genetically rich both for agriculture and from a nutritional point of view; it is harvestable even in adverse conditions, like poor soil quality or hot and humid environments, and has protection from pesticides and pathogens (Katoch, 2013).

 DOI: 10.1201/9781003153481-4

Rice bean is used by farmers to enhance the soil quality, provide weed management and present nutrient-rich food. The nitrogen fixation capability of the crop helps to improve soil quality. The stems of the rice bean plant serve as animal feed; the leaves, for instance, are healthy for chickens. The plant also acts as effective erosion control when grown on slopes and destroys weeds due to lack of air.

The rice bean plant comes in two types of growth patterns: "climbing" and "bush". It can be grown as an annual or a perennial crop. The bush height ranges from 0.98 to 2.46 feet. The plant's yellow flowers mature into spherical pods in clusters of 5–12 around four inches in height. The plants are regarded as short-day plants because they only need a very short amount of day time to undergo flowering. Seeds can be efficiently harvested 90–120 days after plantation, depending time of plantation. The seed is equivalent in size to a rice grain, which is why the plant is referred to as "rice bean". While the plant grows efficiently in hot and humid as well as hot desert conditions, it is harvested much more poorly under wet environmental conditions. Rice bean farming takes place after the cultivation of rice in order to supplement the soil with humus and nitrogen. It can also be cultivated as a rotational crop along with millet and maize. The seeds can be easily cultivated from these high-yield crops as they open when it becomes dry. The pods undergo maturation at a particular time; therefore, one-time cultivation is needed to harvest all seeds. As there is a chance that the dry pods can burst, harvested is recommended as taking place during the morning, when moisture content is higher. The seeds, when preserved for a prolonged period, also possess the ability to undergo efficient germination. Germination efficiency can be increased with a presoaking treatment (Echo Development Notes, issue 83, 2004).

4.3 PESTS, DISEASES AND THEIR MANAGEMENT

Work in 2003 has proved the resistance of the rice bean plant against the bruchid species. In particular, rice bean was found to be extremely resistant to *C. chinesis*, *C. maculatus* and *C. analis*. The study found that the aforementioned bruchid species died after they entered the rice bean seed coat. The work further suggests that the degree of resistance is comparatively wide-spread across wild-type rice bean than cultivated rice bean (Kashiwaba et al., 2003).

Another study carried out in 2007 shows the potential of rice bean in providing resistance against *Callosobruchus maculatus* F. This beetle is responsible for destroying the stored grain, but rice bean was able to inhibit the growth of *Callosobruchus maculatus* at the grub phase. Further examination revealed that the hardness and thickness of the seeds were not the contributing factors in providing resistance; rather, the demise of the grubs was caused by the presence of various biomolecules in cotyledon. Trypsin inhibitors were found to be three to five times higher in resistant rice bean varieties than in vulnerable ones. Further, cysteine protease inhibitors were seven times higher in resistant rice bean varieties than in vulnerable ones (Srinivasan & Durairaj, 2007).

4.4 FUNDAMENTALS OF GENOMICS-MEDIATED CROP IMPROVEMENT

In order to meet the requirement of supplying nutritious food in sufficient quantities to an exponentially increasing human population, a strategy must be developed to address crop improvement. This can be attained by carrying out genetic-level manipulations of genome sequencing and domain assignment, leading to the identification of specific regions in the genome responsible for the translation of nutritional biomolecules and disease resistance. This work can be further expanded toward profiling a crop's available cultivars to look for variation in the regions associated with better yield, pathogen resistance and nutritionally enriched bioactive molecules. In addition, breeding strategies can be brought into action to develop superior cultivars by considering genetic information and hence producing high yield, nutritious and disease-resistant varieties.

In 2005, genomics-assisted breeding (GAB) was introduced to the scientific community to speed up the crop improvement process (Varshney et al., 2005). The general workflow of the GAB is presented in Figure 4.1. First-rate rice crop genome sequencing data was cracked during the introduction of a GAB strategy, and the rice genome became the first crop whose genome was decoded ("The map-based sequence of the rice genome", 2005). Availability of the genomic sequence revolutionized the associated domain and resulted in development

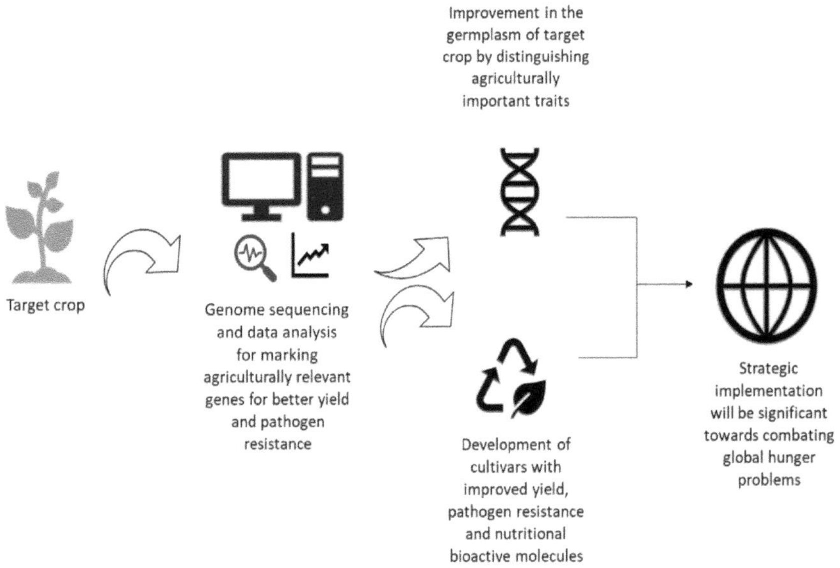

FIGURE 4.1 Generalized workflow of the GAB strategy for the development of new cultivars and improvement of the germplasm quality.

Source: Figure conceptualized from Varshney et al. (2021).

of state-of-the-art genomic resources for retrieval and analysis of the genomic sequence of significance in crop breeding and improvement (Varshney et al., 2021, 2005). Several resources have been developed – e.g., Wheat Genome Variation Database (Wang et al., 2020), MaizeSNPDB (Zhou et al., 2019), BnaGVD (Yan et al., 2021) and SoyKB (Singh et al., 2021) – that contain the genome annotation information for wheat, maize, rapeseed and soybean crops (Varshney et al., 2021). Comprehensive information with regard to exploring omics tools and databases for GAB can be found in our previous studies (Aglawe et al., 2021; Kaul et al., 2019). GAB strategy has played a crucial role in the profiling of allelic-level variation in the genomic regions responsible for expressing beneficial agriculture traits, which in turn have contributed effectively to cultivar development and enrichment of germplasm (Varshney et al., 2021). In the upcoming sections, we assess the impact of genomic tools and technologies in decoding the rice bean crop genome.

4.5 GENOMICS, PROTEOMICS AND GENETICS STUDIES ON RICE BEAN

The whole genome sequence of rice bean was recently decoded employing PacBio and Illumina dais for next generation sequencing (Varshney et al., 2005). Following next generation sequencing topology, a sum of 15521 scaffolds were generated, out of which 31276 genes were predicted using the MAKER pipeline ("The map-based sequence of the rice genome", 2005) with 96.08% functional coverage at 30X coverage data. The gene prediction was carried using the MAKER pipeline, which mediates by repeat identification and subsequent transcriptome and proteome alignment respective of a genome. The sequence library was prepared with an insert size of 350 base pairs and 550 base pairs. The sequence library with 350 a base pair insert size was sequenced using Illumina sequencer, and approximately 30 giga base pair of unprocessed data was obtained. Sequencing of the 550 base pair insert size generated approximately 15 giga base pair of unprocessed data (Varshney et al., 2005).

The annotation process was carried out using the Geneious Prime (version 2019.1) tool, which offers multiple annotation and processing features. Low-quality nucleotides and reads with fewer than 10 base pairs were cut off with Geneious Prime's BBDuk module (Varshney et al., 2021, 2005). Further, the de novo assembly was carried out using the Geneious denovo assembler.

Table 4.1 lists the bioinformatics tools utilized to decode the rice bean genome along with their functions.

4.6 RICE BEAN GENOME SEQUENCING ATTRIBUTES

The preprocessing of the whole rice bean genome gives 69.32 giga bases of the unprocessed data. With the help of Allpath LG assembler, a total 414 Mbps reads were assembled (Schattner et al., 2005). Utilization of PacBio resources, gap closing and post-polishing generated 17,601 scaffolds. The whole processing resulted in 15,521 scaffolds and 36,461 contigs (Varshney et al., 2005). The sum of the scaffolds for N50, N75 and N90 metrics were 7,652, 12,330 and 12,767 respectively (Varshney et al., 2005). Overall, the highest scaffold size for the assembly was 1043.617 kilo base pairs and the lowest scaffold size was 0.886 kilo base pairs (Varshney et al., 2005). More than 7343 scaffolds had a length of more than 10000 base pairs; this increases the possibility of anticipating evolutionary relationships and gene conservation among comparative species (Beier et al., 2017). Out of all the scaffold assembly, 31276 genes were anticipated, and the highest number of base pairs in a single gene were 15628 base pairs (Varshney et al., 2005). Comparative functional genomic analysis revealed 80% of annotated genes common with *Vigna angularis.* Out of all the anticipated genes for rice bean, 19640 genes in mungbean, 17989 genes in cow pea and 16892 genes in adzukibean had orthologous loci respective of rice bean (Varshney et al., 2005).

TABLE 4.1
List of the Bioinformatic Tools Utilized to Completely Annotate the Rice Bean Genome

Computational Tool	Function	Reference
MAKER pipeline	Identification of translational genes.	[6]
Repeat Masker	Masking of the repetitive regions.	[8]
SNAP (Semi-HMM-Based Nucleic Acid Parser)	Gene anticipation.	[9]
AUGUSTUS	Gene anticipation.	[10]
Geneious Prime	Annotation of the genome assembly.	[11]
BBDuk	Cutting of the paired reads.	[7]
MAFFT	Sequence alignment.	[12]
tRNAscan-SE	Classification of selenocysteine tRNA mediated genes.	[13]
MISA (Microsatellite Analysis)	Classification of SSR (simple sequence repeats) markers.	[14]
BBH (Bidirectional Best Hit)	Allotment of KEGG-Orthology terms.	[15]
KAAS (KEGG Automatic Annotation Server)	Pathway examination of the sequence assembly.	[16]
Blast2GO	Annotation of gene ontology.	[17]
BLAST2GO PRO	Annotation of functional genome set.	[18]
Circa	Graphical representation and interaction analysis of genomic datasets.	[19]
Mauve Aligner	Alignment of gene library with reference genomes.	[20]
PlasMapper	Creation and annotation of plasmid maps.	[21]

Biased codon usage analysis with the help of tRNAscan-SE revealed that 577 tRNA operons were found at intron spacer and bound areas (Varshney et al., 2005). Further, a maximum amount of arginine-leucine-serine was observed out of all the anticipated codons (Varshney et al., 2005). Strikingly, introns were present in those tRNA genes that code for specific amino acids, namely leucine, tyrosine, lysine, methionine, asparagine, serine and alanine. The intronic region lengths of leucine, tyrosine, lysine, methionine, asparagine, serine and alanine were 4, 20, 19, 20, 14, 10 and 6 base pairs respectively (Varshney et al., 2005).

Gene ontology terms were assigned to a sum of 23907 dependent genes, out of which the genes anticipated from the rice bean genome assembly showed 11204 roles associated with biological processes (Varshney et al., 2005). Among all the roles/functions associated with biological processes, 6957 corresponded to ion-dependent functions, with supremacy of hemostasis and membrane transport (Varshney et al., 2005). After this, 1560 roles were associated with biological processes viz. DNA replication, transcription, translation, signal transduction, repair mechanisms, duplex unwinding and endonucleolysis (Varshney et al., 2005). Apart from this, 5347 cellular components were recognized as associated with membrane-explicit components (Varshney et al., 2005). Those membrane-explicit components were further subjected to their subcellular location identification on the basis of function (Varshney et al., 2005). A sum of 1534 identified components belong to the nucleus, and the rest of the cellular components were found to be associated with various subcellular locations like golgi, cytoplasm, chloroplast, ribosome and vacuole (Varshney et al., 2005). Identification of the scaffolds associated with various enzymes and other factors, along with the scaffold distribution in number of genes, is presented in Table 4.2 (Varshney et al., 2005).

TABLE 4.2
Distribution Tally of Various Enzyme Classes and Other Factors at the Scaffold and Gene Level of the Rice Bean Genome Assembly

Enzyme Class/ Functions	Total Number of Scaffolds	Total Number of Genes Having the Scaffold Distribution
Isomerases	2	19
Lyases	5	8
Hydrolases	41	683
Transferases	59	1017
Oxidoreductases	64	8550
Activation and Inhibition	40	855
Dogmatic	7	88
Channel	13	87
Transporter	38	522
Binding	194	4189

TABLE 4.3
Identification of the Various Biological Process's Arrangements in the Rice Bean Genome and Its Functional Correspondence with KEGG-Orthology

Biological Processes (Number of Arrangements/ Processes)	Number of Orthologs Identified by KEGG-Orthology
Genetic information processing arrangements (4)	22
Environment information processing arrangements (3)	40
Cellular processing arrangements (5)	33
Organism system arrangements (10)	87
Human disease arrangements (12)	89

*Biological processes are assumed to be in the form of an arrangement.

A sum of 515 functionally active orthologs was expanded from 19173 lab-validated KEGG-gene related pathways. A total of 54 functionally enriched prevalent arrangements were co-related with KEGG-Orthology accessions. A list of all functionally enriched arrangements corresponding to biological processes and KEGG-Orthology is shown in Table 4.3 (Varshney et al., 2005).

The majority of any crop's vital genes usually code for transcription factors (Aoki-Kinoshita & Kanehisa, 2007) Gene expression in plants is regulated by various factors, such as transcription factors, chromatin regulators and transcription regulators (Masoudi-Nejad et al., 2007). With the help of the Hidden Markov Model, various factors, including those that regulate growth and membrane transport of the rice bean, were disclosed (Varshney et al., 2005). For the rice bean genome assembly, a total of 11202 transcription factors were recognized (Varshney et al., 2005). From the rice bean genome assembly, out of all anticipated genes, a total of 15 gene clusters of transcription factors with vital functions were recognized. Table 4.4 shows the identified gene clusters along with the transcription factors (Varshney et al., 2005).

Identification of the genomic regions suitable for genome editing paves the way to get to the desired output (Varshney et al., 2005). In case of rice bean, 33 genes distributed over 7049 loci were distinguished as having the vector specificity (Varshney et al., 2005). Nine genes were distinguished as localization genes, with an abundance of MAPKK_NES gene (Varshney et al., 2005). Out of all the anticipated genes, 20 promoter-specific genes were identified, out of which lpp promoter was found the most, at 6225 genomic regions (Varshney et al., 2005). Moreover, 31 regulatory elements were identified as present at 30939 loci, out of which TE5pUTR (Translational_Enhancer_5_prime_UTR) was present at 3105 loci of the rice bean genome (Varshney et al., 2005). Finally, 11 termination sites were distinguished at 21996 loci, out of which synthetic polyadenylation signal terminator was identified at 5903 loci (Varshney et al., 2005).

TABLE 4.4
Various Transcription Factors Along with Their Gene Names that Are Found in the Rice Bean Genome Assembly

Gene Cluster Name	Transcription Factor	Gene Cluster Name	Transcription Factor
Nitrilase Family Member 2	NIT2	GT2 facilitative glucose transporter GT2	GT-2
Nuclear Factor	SEF4	Hexamer motif 5'-ACGTCA-3'	HBP-1a(c14)
TATA-Box binding protein associated factor 1	TAF-1	Opaque endosperm 2	Opaque-2
60K Protein	60K Protein	Sequence-specific single stranded DNA binding proteins-2	SsDBP-2
Putative chromosomal passenger protein	CPC1	Putative homeodomain-like transcription factor superfamily protein	Zmhox1a
G-Box binding factor	GBF	TEA/ATTS transcription factor	abaA
GT1 facilitative glucose transporter GT1	GT-1	Common Plant Regulatory Factor 1-like	CPRF-1

4.7 FUTURE PERSPECTIVES

This chapter clearly depicts the importance of the rice bean crop as a potential nutritional enrichment crop to feed a hungry world. Identification of various regions that are suitable for genetic engineering will pave the way to producing crops/strains of rice bean that will be genetically modified/altered in order to respond/grow according to particular environmental conditions. Further advancements can be made in the rice bean crop to produce hybrid rice bean strains with fortified nutritional components.

4.8 DISCUSSION AND CONCLUSIONS

Rice bean is an underutilized crop whose potential nutritional and economic benefits are not known to farmers in developing and underdeveloped countries. Rice bean has many nutritional components, including proteins, vitamins, starch, minerals and oils. Whole genome sequencing was a key milestone in rice bean research and has paved the way for identification of regions responsible for disease resistance as well as other regions of the genome (e.g., vector specificity, promoters, regulatory elements and terminators) that can be utilized to produce hybrid strains with the help of genetic engineering. Governing authorities from the concerned departments should come up with strategies in developing and underdeveloped countries to make farmers aware of the benefits of rice bean and to encourage them to harvest this crop for the welfare of the underprivileged.

REFERENCES

Aglawe, S. B., Verma, A. K., & Upadhyay, A. K. (2021). Bioinformatics tools and databases for genomics-assisted breeding and population genetics of plants: A review. *Current Bioinformatics*, *16*(6), 766–773.

Aoki-Kinoshita, K. F., & Kanehisa, M. (2007). Gene annotation and pathway mapping in KEGG. *Methods in Molecular Biology*, *396*, 71–91. https://doi.org/10.1007/978-1-59745-515-2_6

Beier, S., Thiel, T., Münch, T., Scholz, U., & Mascher, M. (2017). MISA-web: A web server for microsatellite prediction. *Bioinformatics*, *33*(16), 2583–2585. https://doi.org/10.1093/bioinformatics/btx198

Kashiwaba, K., Tomooka, N., Kaga, A., Han, O. K., & Vaughan, D. A. (2003). Characterization of resistance to three bruchid species (Callosobruchus spp., Coleoptera, Bruchidae) in cultivated rice bean (Vigna umbellata). *Journal of Economic Entomology*, *96*(1), 207–213. https://doi.org/10.1093/jee/96.1.207

Katoch, R. (2013). Nutritional potential of rice bean (Vigna umbellata): An underutilized legume. *Journal of Food Science*, *78*(1), C8–C16. https://doi.org/10.1111/j.1750-3841.2012.02989.x

Kaul, T., Eswaran, M., Thangaraj, A., Meyyazhagan, A., Nehra, M., Raman, N. M., Bharti, J., Badapanda, C., & Balamurali, B. (2019). Rice Bean (Vigna umbellata) draft genome sequence: Unravelling the late flowering and unpalatability related genomic resources for efficient domestication of this underutilized crop. *bioRxiv*, 816595.

Lolas, G. M., & Markakis, P. (1975). Phytic acid and other phosphorus compounds of beans (Phaseolus vulgaris L.). *Journal of Agricultural and Food Chemistry*, *23*(1), 13–15.

The map-based sequence of the rice genome. (2005). *Nature*, *436*(7052), 793–800. https://doi.org/10.1038/nature03895

Masoudi-Nejad, A., Goto, S., Jauregui, R., Ito, M., Kawashima, S., Moriya, Y., Endo, T. R., & Kanehisa, M. (2007). EGENES: Transcriptome-based plant database of genes with metabolic pathway information and expressed sequence tag indices in KEGG. *Plant Physiology*, *144*(2), 857–866.

Schattner, P., Brooks, A. N., & Lowe, T. M. (2005). The tRNAscan-SE, snoscan and snoGPS web servers for the detection of tRNAs and snoRNAs. *Nucleic Acids Research*, *33*(Web Server issue), W686–W689. https://doi.org/10.1093/nar/gki366

Singh, B., Mohanty, D., Bakshi, V., Gujjar, R. S., & Upadhyay, A. K. (2021). The distinction of omics in amelioration of food crops nutritional value. In *Bioinformatics for agriculture: High-throughput approaches* (pp. 85–99). Springer.

Srinivasan, T., & Durairaj, C. (2007). Biochemical basis of resistance to in rice bean Vigna umbellata Thunb.(Ohwi and Ohashi) against Callasobruchus maculatus F. *Journal of Entomology*, *4*(5), 371–378.

Varshney, R. K., Bohra, A., Yu, J., Graner, A., Zhang, Q., & Sorrells, M. E. (2021). Designing future crops: Genomics-assisted breeding comes of age. *Trends in Plant Science*, *26*(6), 631–649. https://doi.org/10.1016/j.tplants.2021.03.010

Varshney, R. K., Graner, A., & Sorrells, M. E. (2005). Genomics-assisted breeding for crop improvement. *Trends in Plant Science*, *10*(12), 621–630. https://doi.org/10.1016/j.tplants.2005.10.004

Wang, J., Fu, W., Wang, R., Hu, D., Cheng, H., Zhao, J., Jiang, Y., & Kang, Z. (2020). WGVD: An integrated web-database for wheat genome variation and selective signatures. *Database (Oxford)*. https://doi.org/10.1093/database/baaa090

Yan, T., Yao, Y., Wu, D., & Jiang, L. (2021). BnaGVD: A genomic variation database of rapeseed (Brassica napus). *Plant & Cell Physiology*. https://doi.org/10.1093/pcp/pcaa169

Zhou, W., Wang, L., Zheng, W., & Yao, W. (2019). MaizeSNPDB: A comprehensive database for efficient retrieve and analysis of SNPs among 1210 maize lines. *Computational and Structural Biotechnology Journal*, *17*, 1377–1383. https://doi.org/10.1016/j.csbj.2019.10.003

5 Omics Technologies Unravelling the Plant–Pathogen Interaction and Stress Response

Sakthi Ambothi Rathnasamy, Sellamuthu Gothandapani, Selvi Chellamuthu, Amrita Chakraborty, Dhandapani Gurusamy, and Amit Roy

5.1 INTRODUCTION

Various physiological and chemical processes occur in plants to absorb and convert nutrients and water into a usable form essential for plant growth. In addition to those processes, plants must handle mechanical, biotic (insect, pathogen), and abiotic (drought, heat, and salinity) stress factors. Adaptation of plants depends on their ability to quickly perceive insect/pathogen attacks and changes in the environment to express an innate adaptive response. Plants respond to these environmental changes by initiating several physiological reactions, leading to the synthesis of a wide array of secondary metabolites.

The response of plants to insect attack, pathogen infection, and abiotic stresses like drought, soil flooding, and high or low temperatures have been gradually unravelled. Plant pathogens are generally divided into nectrophs (pathogens that destroy host cells and feed on the contents) and biotrophs (organisms that derive nutrients from living tissue for their survival) according to their living nature (Glazebrook et al., 2005). Hemibiotrophs possess both types, depending on the stage of their lifecycle. Plants develop intricate mechanisms to perceive pathogen attacks and convert these perceptions into immune responses against their attacker. Plants initially recognize microbial pathogens using common features like flagellin, glycoproteins, chitin, and polysaccharides, referred to as pathogen-associated molecular patterns (PAMPs) (Jones & Dangl, 2006; Chisholm et al., 2006; Göhre & Robatzek, 2008; Nurnberger & Kemmerling, 2006). Plant immunity responds by activating several downstream signalling processes through pattern recognition receptors (PRRs) that activate PAMPs. These responses occurring in plants are modulated by various networks of signalling pathways induced by elicitors/number of small molecules,

 DOI: 10.1201/9781003153481-5

phytohormones (Pieterse et al., 2012; De Vleesschauwer et al., 2014; Kazan, 2015), Ca2+ signalling (Seybold et al., 2014), and reactive oxygen species (Wang et al., 2013; Baxter et al., 2014). Among those signalling molecules, hormones and hormonal cross-talk are important in optimizing plant responses to biotic and abiotic stresses commonly encountered in the environment (Bostock et al., 2005; Pieterse & Dicke, 2007).

Interaction and cross-talk among phytohormone-mediated signalling pathways depend on the nature of stresses. The primary biochemical response mechanisms induced under various stresses are jasmonic acid (JA) and salicylic acid (SA) signalling pathways. Identification of molecular factors and signalling mechanisms involved in the interactions of phytohormones and understanding the cross-talk between different signalling pathways will provide information on the phytohormone-mediated defence signalling network in plants. These molecular factors and mechanisms underlying pathogen-related stress were unravelled using several molecular omics technologies.

Emerging technologies such as "omics" are collectively used to recognize the influence of plant cells and tissues at the molecular level by applying current technologies like genomics, proteomics, transcriptomics, metabolomics, system biology, and bioinformatics. Recent advances in molecular biology approaches have increased the detection of pathogens (novel and reported) and mechanisms underlying the interaction between plants and pathogens. Among various molecular methods, omics tools allow comprehensive examination and characterization of plant and microbial interaction (Crandall et al., 2020). Ecological invasions (Gilbert et al., 2010) and climate change modulate microbes and their environment (Santini et al., 2015). The overall development in omics approaches to plant disease ecology is essential in a rapidly changing environment and evolving biotic/abiotic stress factors according to changing environmental factors. Two or more omics technologies can be combined and applied to understand the basic functional principles of cellular function (Zhang et al., 2010; Assche et al., 2015). Knowledge of genetics and its approaches can be linked to omics technologies to explore various processes in plant disease ecology. Novel methodologies have significantly improved disease resistance to pests and diseases by transferring genes governing resistance to susceptible varieties (Coleman-Derr & Tringe, 2014). This review discusses the different omics technologies used to unravel the genes and factors responsible for disease resistance and its application in plant–pathogen interaction.

5.2 OMICS: A NEW AREA OF HIGH-THROUGHPUT RESEARCH

Plant scientists have made significant progress in detecting changes at the genome level and gene products like transcripts, proteins, and metabolites produced in different tissues through novel high-throughput technologies. These omics technologies, include genomics, transcriptomics, metabolomics, metagenomics, and proteomics (Figure 5.1).

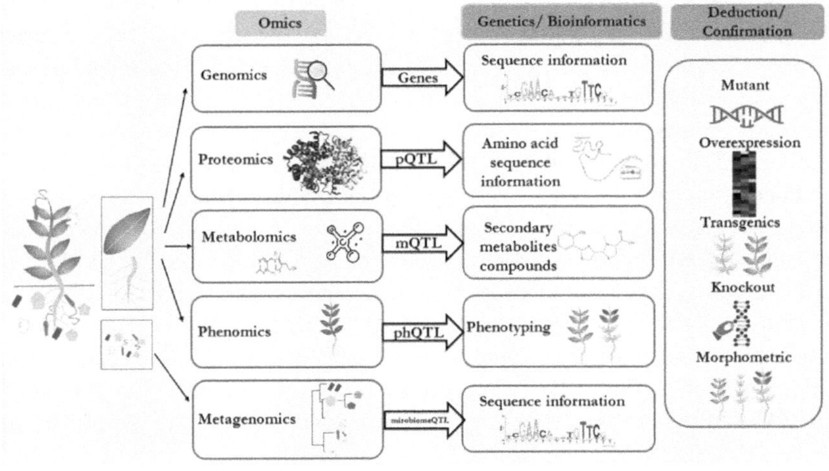

FIGURE 5.1 Multi-omics approaches from the genotype to the phenotype and metagenomics inform plant disease ecology and microbial communities. The identification of the causal genes underlying the QTL, and ultimately their functional characterization, will be facilitated by combined analysis of the data generated using different profiling techniques and additional information obtained using bioinformatics tools.

In molecular biology, the term "-ome" in omics refers to studying the overall or total changes in an organism. Omics refers to the collective molecular biology techniques used to explore the roles and interactions of various biomolecules, gene expression, protein levels, metabolites, or even ionic levels, which are essential for cellular function.

The different omics technologies, including genomics, transcriptomics, proteomics, and metabolomics, study the changes at the genome, transcriptome, protein, and metabolic level in the cell. The aforementioned technologies can generate data corresponding to phenotypic changes and integrate them to an ecological context. Recently, new genomics-based technologies like next-generation sequencing (NGS), gene silencing (RNA interference), and gene editing (CRISPR-Cas9) tools have provided novel information for improving different crop plants to sustain abiotic and biotic stress (Luan et al., 2015).

5.2.1 Genomics: Observing the Organism's Potential at DNA Level

The complete set of DNA content of the organism can be referred to as the genome. Genomics is about whole-genome study (Keller, 2011), including its structure, function, and comparison with the environment and other organisms. The different subsidiaries of genomics include structural genomics, dealing with the structure and organization of genes in a genome; functional genomics, which studies the function of genes and their expression under different environmental conditions; and comparative genomics, dealing with the comparison of genomic features of different organisms to study their evolution pattern (Rubin et al., 2000). The completion of the genome sequencing projects initially in *Arabidopsis* and rice in the early 2000s paved the way for plant genomic research for different traits. Recent advances in DNA marker technologies help detect desirable traits through single nucleotide polymorphism (SNP).

Functional genomics uses DNA sequencing, gene transcription, and protein translation to study the function of gene expression and further translation of transcriptome to the protein on a genome-wide approach. The combination of these omics data has increased understanding of various mechanisms and information related to host–pathogen biology and plant health management (Lindeberg, 2012; Sundin et al., 2016). This could develop new varieties or hybrids tolerant to drought and cold, and resistant to pathogens, increasing yield (Yao et al., 2011; Le et al., 2012; Agarwal et al., 2014).

In the 1970s, RFLP-based molecular markers were applied for crop improvement. Subsequently, PCR-based markers, including RAPD, RFLP, SAMPL, SRAP, TRAP, and DArT, were used for DNA fingerprinting, marker-assisted breeding (MBC), marker-assisted backcross breeding (MABC), and QTL mapping. However, advances in sequencing technologies enabled the identification of SNPs in the whole genome and their downstream utilization for crop improvement. Advanced next-generation sequencing technologies (NGS) using Illumina/Solexa, Ion Torrent Personal Genome Machine (PGM), and Pacific Biosciences (PacBio) sequencing methods revolutionized genomic research in plants (Heather & Chain, 2016). These technologies provide large collections of markers, high-density genetic maps, and high-throughput genotyping of experimental populations, and these technologies can be incorporated into existing breeding methods (Kaya et al., 2013; Sakiyama et al., 2014). Complex plant genomes can be characterized using NGS technologies without a reference genome (Liu et al., 2014). Recent advancements in these technologies provide sequencing of whole genomes at a lesser cost, and the new variants can be identified among a larger population. Also, it aids in estimating the genetic diversity between the germplasms and further usage in breeding programmes (Varshney et al., 2009). Transcriptome generates massive expression datasets of a species under a particular condition, and these data can be used to identify new genes conferring resistance to biotic and abiotic stresses (Cohen & Leach, 2019; Kang et al., 2020). These genome-wide markers were used for linkage mapping or association mapping. Association mapping enables the identification of SNPs in candidate genes regulating complex traits and their association with other traits contributing to the phenotype (genome-wide association mapping, GWAS) (Morrell et al., 2011).

5.2.2 Transcriptomics: Looking at the Expressed Portion of the Genome

The term transcriptomics refers to studying the complete expression profile of genes under specific circumstances in particular plant cells and tissues (Duque et al., 2013). The initial transcriptomic approach started in the 1970s, isolating mRNA from silkmoth and converting it into complementary DNA (cDNA) by reverse transcriptase (Sim et al., 1979). Then Sanger sequencing was used to sequence individual transcripts as expressed sequenced tags (Sutcliffe et al., 1982; Putney et al., 1983). In the 1990s, sequencing-based transcriptomic methods were developed, starting with the serial analysis of gene expression (SAGE) (Velculescu et al., 1995) and digital gene expression analysis (Audic & Claverie, 1997). The SAGE method works by sequencing the RNA as random concatenated fragments, and digital gene expression analysis is a variant of SAGE. However, these methods were entirely replaced by high-throughput sequencing methods such as microarrays and RNA-seq in the mid-1990s and 2000s (Wang et al., 2009). Microarrays comprise short nucleotide probes that are coated. The abundance of the transcript was determined by the hybridization of transcripts labelled with fluorescent probes, and it requires prior information regarding the plants in the form of annotated/reference sequences (Barbulovic-Nad et al., 2006).

RNA-seq refers to the sequencing of complete transcriptomes converted into cDNAs, in which abundance is estimated from the number of counts for each transcript. Earlier RNA-seq work was done using the 454 technology with 105 transcripts in 2006, and it shows sufficient coverage to quantify relative transcript abundance (Bainbridge et al., 2006). After the development of the new technologies such as Solexa/Illumina technologies, the usage of RNA-seq for the transcriptome study increased in extent. It combines high-throughput technologies and computational methods to study entire transcripts and their abundance in an organism (Lappalainen et al., 2013). It requires less RNA concentration than microarray, and it forms computational reconstruction of RNA by aligning sequenced reads with the reference genome. RNA-seq is used to identify novel genes within a genome and information regarding the differentially expressed gene under specific conditions (Figure 5.2).

5.2.3 Proteomics: Looking at the Function of the Genome

The proteome refers to the total expressed protein in the cell, tissue, or given organism under particular conditions (Tyers & Mann, 2003). There are several types of proteins expressed in the plant system, of which whole proteome (common proteins measured under stress conditions) and phosphoproteome (quantification of phosphorylated proteins expressed under a given system) are common proteomes (Helmy et al., 2011). Improving the quantification of protein and determining new proteins in the plant system has paved the way for analysing post-translational modifications and their differential expression at different life stages under biotic and abiotic stress conditions (Xie et al., 2016). In protein expression

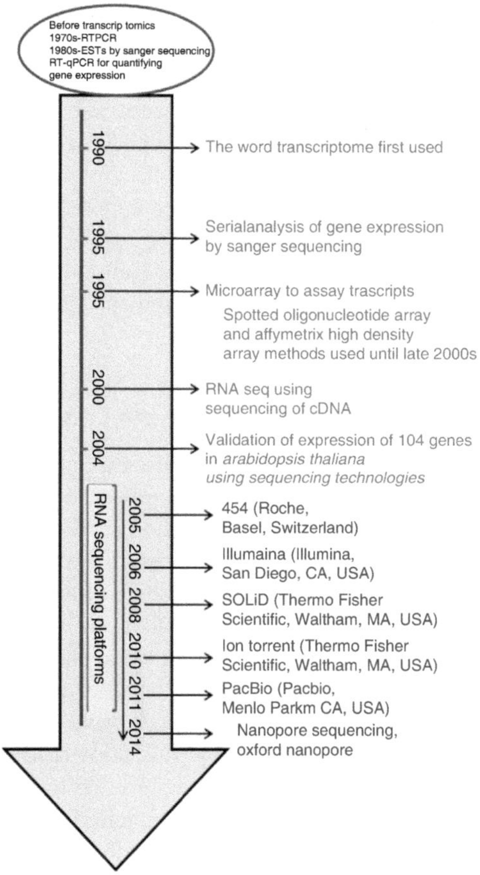

FIGURE 5.2 Development of transcriptomics technology over the years leading to an increased number of plant omics studies.

Source: based on the Web of Science database

studies, until recently, most labs used two-dimensional electrophoresis, and difference gel electrophoresis (DIGE) and other proteomic methods include protein profiling, protein interaction analysis, and protein mapping in plant proteomics (Duque et al., 2013; Komatsu et al., 2014; Arentz et al., 2015; Zandalinas et al., 2018). For protein identification, microcapillary liquid chromatography, coupled with the other technologies used in the proteomic approach, includes matrix-assisted laser desorption/ionization mass spectrometry (MALDI-MS), liquid chromatography mass spectrometry (LC-MS/MS), ion trap mass spectrometry (IT-MS), and latest Q Exactive HF-X Quadrupole-Orbitrap MS System (with uHPLC). The host–pathogen interaction has been determined by two-dimensional gel electrophoresis (2-DE) combined with matrix-assisted laser desorption ionization and tandem time-of-flight mass spectrometry (MALDI-TOF-TOF/MS) (Pentimone et al., 2020).

Nowadays, shotgun analysis (gel-free proteomics) has become the dominant proteomic platform, and it affords deeper proteome coverage in which more proteins were identified than the gel-based approach. The shotgun approach eliminates many pre-processing steps before mass spectrometric analysis of proteins (peptides) compared to the traditional gel-based approach, minimizing the loss of peptides. However, both gel-based and gel-free approaches are complementary, and the former is still a valid approach (Sghaier-Hammami et al., 2020; Lakhneko et al., 2020). Increasing plant genome availability will boost future proteomic research on crops.

5.2.4 Metagenomics: Looking at the Invisible Associates

Metagenomics refers to the identification of disease-causing microscopic organisms. This technique identifies microorganism/disease-causing agents based on phenotypic symptoms, microscopic observation, and preliminary information from previous reports of the particular pathogen (Ward et al., 2004). Initially, microbes are identified based on their morphological appearance, growth index, and biochemical profiles (Oliver et al., 1991; Colwell et al., 1996). Then, in the late 1970s, ribosomal RNA genes were used as molecular markers (Woese & Fox, 1977), and this method, along with Sanger automated sequencing, revolutionized the microbial classification system (Sanger et al., 1977). Similarly, polymerase chain reaction (PCR), fluorescent in situ hybridization (FISH), denaturing gradient gel electrophoresis (DGGE and TGGE), restriction fragment length polymorphism (RFLP), and terminal restriction fragment length polymorphism (T-RFLP) have a significant impact on environmental microbial community and diversity. The metagenomics approach identified new genes, their function, and the metabolic products (Handelsman et al., 1998). The average sequencing length produced by Sanger sequencing is 650 bp, so it cannot be used to sequence the whole genome analysis.

Nevertheless, the next-generation sequencing platforms generate millions of DNA molecules with different sequence lengths and yield high sequence lengths (Glenn, 2014; Sanchez-Flores & Abreu-Goodger, 2014). Pyrosequencing, ion torrent platform, Illumina, and recently PacBio (Fichot & Norman, 2013) or Oxford nanopore sequencing were known NGS sequencing platforms used for the metagenomics approach. Amplicon sequencing and shotgun metagenomics can determine the microbial diversity in a particular environment. Amplicon sequencing involves amplifying a specific conserved region such as 16S rRNA (prokaryotes) and internal transcribed spacers (ITS) (eukaryotes) (Sharpton, 2014). The shotgun metagenomics approach aids in reconstructing the complete genome or large fragments even for newly isolated organisms and characterizes noncoding and coding regions that deliver novel genes. The shotgun approach makes species-level identification of microbes feasible compared to amplicon sequencing, which delivers genus-level information with high confidence. Using this approach, microorganisms belonging to the unclassified group can be identified, which can be used in biofertilizer production or as biocontrol agents. It is also used in identifying microbiomes associated with resistant plants, and these microorganisms can be isolated and applied to susceptible plants to suppress symptoms (Kwak et al., 2018). Recent advancements in sequencing approaches such as metatranscriptomics or metaproteomics can selectively identify the microbial gene or protein expression inside the host. For instance, using the metatranscriptomics approach, the gene expressions were identified in the bacterial population colonized in the root gall of root-knot nematode (Tian et al., 2015)

5.2.5 Metabolomics

Metabolomics is an omics technology, studying the presence of particular metabolites and quantifying all metabolites in a living organism. Oliver et al. (1998) introduced the term metabolomics to substantiate the change in metabolite concentration according to environmental conditions. Due to the diversity of chemical structures inherent in the metabolome, no single analytical method can analyse these structures simultaneously. Metabolomics was applied for analyses of important compounds using GC-MS (gas chromatography and mass spectrometry) (Sumner et al., 2003). *In-planta* biological system applications of profiling techniques began in the 1990s to investigate the mode of action of herbicide molecules (Sauter et al., 1991). Plants are sessile in nature, and they produce various metabolites under abiotic and biotic stress factors. Identifying these stress-induced metabolites or the compounds synthesized for plant survival and genomic and phenotypic data provides a better understanding of plants' genetic improvement (Badjakov et al., 2012). Since each metabolite has different chemical and physical properties, analytical and separation techniques are used to obtain the metabolic profile of a cell/tissues of the plant or given sample (Jogaiah et al., 2013).

Nuclear magnetic resonance (NMR) and mass spectrometry (MS) are two main metabolomics techniques that led to the identification and characterization of hundreds of metabolites in plants (Catola et al., 2016). To determine the expression level of metabolites and their corresponding gene/protein, metabolomics is usually used with other omics technologies like transcriptomics or proteomics (Srivastava et al., 2013). NMR spectroscopy aids in obtaining the molecular structure of molecules (metabolites) and novel components. Ultra-high performance liquid chromatograph (UHPLC) is used for metabolite detection, and UHPLC coupled to a quadrupole time-of-flight high-definition mass spectrometer (qTOF HD-MS) system distinguishes metabolite classes of different polarities. In maize, an examination of the multifactorial nature of mycotoxin resistance and plant–pathogen interaction mechanisms found changes in lipid metabolites related to fumonisin accumulation under open field conditions (Pentimone et al., 2020; Righetti et al., 2021). The change in metabolite concentration to stress response in a model plant like *Arabidopsis* was done using metabolomics (Cook et al., 2004) and also in food crops such as cereals (Sicher & Barnaby, 2012), common bean (*Phaseolus vulgaris*) (Broughton et al., 2003), and poplars (*Populus* x *canescens*) (Behnke et al., 2010).

5.2.6 Phenomics

Phenomics studies the set of phenotypic characters produced by an organism over its growth period and varying environmental conditions (Houle et al., 2010). Genetic improvement of crops is the primary objective of improving crop productivity to meet sustainable food grain production. Molecular breeding strategies need phenotypic data along with genotypic information (Jannink et al., 2010). Molecular breeding populations possess a minimum of 200 lines and a maximum of 10,000 lines, and phenotyping of huge populations/lines is a cumbersome process. To overcome this bottleneck, high-throughput and automated phenotyping technology is needed to characterise phenotypic traits with maximum accuracy. Plant phenomics emerged as an essential tool for gathering more phenotypic data at different crop stages, from a single cell to a population level (Houle et al., 2010; Lobos et al., 2017). Crop phenomics integrates various fields, including life science, agronomy, information science, and engineering and combines scientific knowledge with artificial intelligence to collect different phenotypic characters in complex environments (Zhao et al., 2019). The most commonly measured phenotypic data, such as leaf area, length, and fruit volume, can be phenotyped easily (Klukas et al., 2014; Zhang et al., 2017a). A smartphone app platform was developed for phenotyping at the organ level, and it can be measured by taking pictures and analysing the images (Confalonieri et al., 2017). Two-dimensional cameras were integrated into phenotyping platforms that record the phenotype of branched-type plants and track the plant's growth at the organ level (Brichet et al., 2017).

Automated phenotyping platforms can be applied to controlled and field environments. High-throughput plant phenotyping platforms can be fixed to a growth chamber, or greenhouse in a controlled environment, where either the plants can be moved to an imaging system or an imaging system moves on the plant canopy to record the observations (Fiorani & Schurr, 2013; Virlet et al., 2016). Phenotyping at the field level is a cumbersome process, and it is a crucial component of genetic improvement since the overall effect of interaction between genetic factors with environmental stress factors was measured (Araus & Cairns, 2014). The field-level phenotyping platforms use ground wheeled or aerial vehicles, sensors, cameras, and high-performance computing to record the spatial and temporal data throughout the crop cycle (Fritsche-Neto & Borém, 2015). Recently, human-crewed aircraft or automated aerial vehicles have been used for crop phenotyping with maximum resolution (Liebisch et al., 2015). The sensors carried by these phenotyping platforms include infrared thermal imagers, multispectral cameras, light detection and ranging (LIDAR), and hyperspectral sensors. These advanced tools are applied for canopy modelling, estimating crop biomass, and monitoring physiological status.

5.3 ROLE OF MULTI-OMICS IN PLANT DEFENCE

The use of omics technologies contributed to studying plants' responses to biotic and abiotic stress factors, enhancing the quality of food crops (Zhang et al., 2016). The use of omics approaches improved consistency and predictability in breeding crops and reduced the time for producing better quality crops (Van Emon, 2016). By combining the system biology and omics approaches, the pathogen–host interaction database (PHI-base) was created in 2005 to verify pathogenicity data and virulence data of pathogens infecting the crop plants (Winnenburg et al., 2008).

On infection of plants, pathogens interact with host proteins to suppress the immune system. Plants express disease-resistant proteins in response to pathogen attacks and trigger defence response. The specific transcripts were identified using transcriptome sequencing to detect the resistance mechanism of tomato plants in response to *Pseudomonas syringae* (Pombo et al., 2014). The alteration in genes and metabolites was observed in tomato infected with tomato leaf curl virus. Increased synthesis of polyamines and phenolic and indolic metabolites led to defence-related compounds (Sade et al., 2015). The molecular mechanisms underlying the interaction of wheat and Arbuscular mycorrhizal fungi were elucidated using transcriptomic and protein profiling techniques. It identifies the metabolic pathway involved in synthesizing metabolic compounds that enhance plant biomass, mineral nutrition, and a bio-protective effect against the leaf pathogen *Xanthomonas translucens* (Fiorilli et al., 2018). The genes chalcone synthase (CHS) and chalcone isomerase (CHI) involved in the flavonoid pathway contribute to disease resistance and the expression pattern of these genes in resistant grapevine cultivars (Gao et al., 2012). The transcriptomics study revealed the expression of differentially expressed genes related to flavonoid synthesis, including chalcone synthase and phenylalanine ammonia-lyase after inoculation of pathogen causing grey blight in tea (Wang et al., 2021).

Several antimicrobial proteins expressed during pathogen infection were identified through proteomic approaches (Mehta et al., 2008). The expressed proteins with antimicrobial properties were identified, and information regarding these peptides was listed in the PhytAMP database (www.phytamp.pfba-lab-tun.org) (Hammami et al., 2009) and CAMP (Collection of Anti-Microbial Peptides, www.camp.bicnirrh.res.in) (Thomas et al., 2010). Searching databases identified the disease resistance genes (R genes). The R genes conferring resistance to blackleg, sclerotiana, and clubroot resistance were identified by searching details of protein motifs in the databases (Stotz et al., 2018).

The resistant barley cultivar synthesizes nearly 500 metabolites resistant to Fusarium head blight, one of the common diseases in barley, maize, and wheat (Bollina et al., 2010). NGS-based SNP genotyping has been used for QTL mapping of disease resistance traits, and genome-wide association studies were used to identify candidate genes. Using this approach, several significant QTLs for sclerotinia quantitative resistance were introgressed from *Brassica oleracea* into *Brassica napus* possessing superior seed yield and quality (Mei et al., 2020). An essential application of metagenomics is the recovery of bacterial phyla and their functionality from symbiotic relationships. Intimate associations exist between various eukaryotic hosts in the marine environment and phylogenetically distinct microbial communities.

5.4 PRELUDE ON THE INTERACTION OF PLANT AND PATHOGEN

The interaction between higher plants and microbial pathogens such as viruses, bacteria, and fungi leads to plant diseases and the most complex phenomena in biology. Understanding plant–pathogen interactions is critical to maintaining and improving plant health (Figure 5.3) (Rodriguez et al., 2019). Broadly, plant–pathogen interactions can be viewed: microbial pathogenesis, cross-talk between pathogenesis and plant immunity, and resistance-derived molecules, mainly proteins, sugars, and lipopolysaccharides (Schneider & Collmer, 2010; Boyd et al., 2013). Plant-induced responses determine these plant-mediated effects: damage by pathogens that involve changes in physical adaptation (spines, trichomes) and chemical defences (metabolites). Plant signalling pathways involved in up-regulating these defences (salicylic acid, jasmonic acid, ethylene pathway) lead to predictable outcomes of plant-mediated interactions among attackers (Mithöfer & Boland, 2012; Núñez-Farfán et al., 2007; Moreira et al., 2018). Similarly, plants

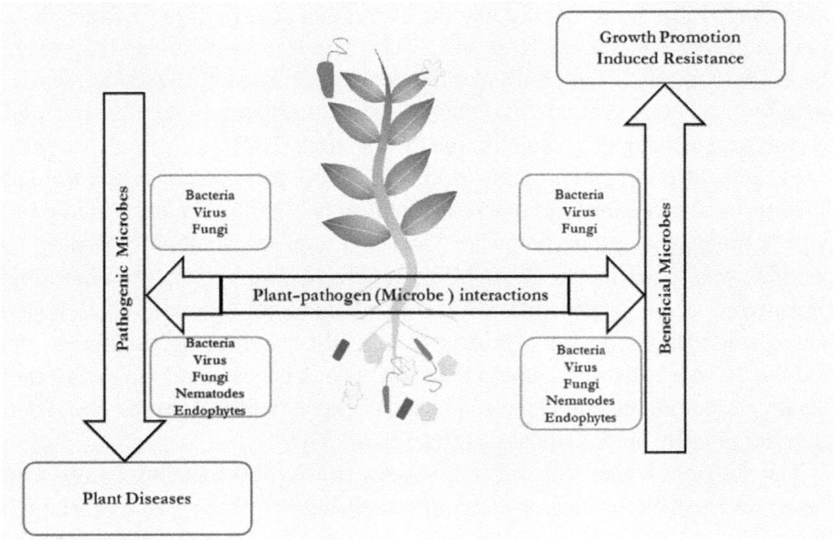

FIGURE 5.3 Plant–pathogen interactions. This model explains the complex network of interactions occurring between the plant and pathogens/beneficial microbes. The beneficial microbes promote plant growth and/or suppress plant diseases via a variety of mechanisms, which include improved nutrient acquisition, production of growth regulators, and biosynthesis of pathogen-inhibiting compounds (secondary metabolites).

induce immunity through growth regulators, such as auxins, brassinosteroids (BRs), cytokinins (CKs), abscisic acid (ABA), and gibberellins (GAs), that contribute to plant immunity (Berens et al., 2017; Islam et al., 2019). Plant defence pathways that mediate plant microbial interactions have emerged as a broadly used methodology for omics techniques. An increasing number of studies started using multiple omics methodologies on the same samples to enhance our understanding of biological systems (Altelaar et al., 2013; Peyraud et al., 2017).

5.4.1 Salicylic Acid Signalling Pathway and Defences

Salicylic acid (SA) is one of the phenolic compounds synthesized by plants and is involved in various plant processes, including development, flowering, senescence, and responses against abiotic and biotic stress. SA is extensively reported for its role in local and systemic acquired resistance (LAR and SAR) against pathogens (Dempsey et al., 2011; Wenig et al., 2019; Lefevere et al., 2020). Subsequently, it was involved in many vital processes, including lignin and pigment biosynthesis regulation of responses to abiotic stresses (Han & Kahmann, 2019; Yu et al., 2020). Also, SA regulates plant photosynthetic parameters, including stomatal closure, influencing the structure of chloroplasts and enzymes involved in photosynthesis (Kohli et al., 2017). SA induces essential plant defence components through SA binding proteins (SABPs) that signal downstream responses to the non-expressor of PR1 protein (NPR1) pathway (Gao et al., 2015). Non-expressor of PR1 protein (NPR1) is an essential component in plant defence signalling receptors and may promote immune signalling by relieving sequestration of NPR1's trans-activating domain via NPR1-dependent and NPR1-independent pathways (Wu et al., 2012). The NPR1-dependent pathway is best understood the importance of signalling, which reported Arabidopsis NPR1 mutants failed to activate immunity following SA treatment or pathogen infection (Gao et al., 2015; Dempsey & Klessig, 2017). Under unchallenged conditions, NPR1 is predominantly sequestered in the cytosol as a higher molecular oligomeric complex. However, in response to pathogen attack, SA induces the conversion of NPR1 oligomeric protein to a monomeric and is transported to the nucleus to activate PR gene expression (Mou et al., 2003; Lee et al., 2015). NPR1 regulates the expression of immune-associated genes through interaction with several TGA transcription factors (Mou et al., 2003; Dempsey & Klessig, 2017).

Pathogenesis-related (PR) proteins (PR1, 4 & 5 – anti-oomycete and antifungal; PR2 – β-1,3-glucanases; PR3 – chitinases; PR6 – proteinase inhibitors) are diverse molecules that accumulate in response to pathogen attachment to prevent biotic stress (Pajerowska-Mukhtar et al., 2013). PR-1 gene expression has been primarily considered as a SA marker due to its strong induction during SA-mediated plant immune responses. PR-1 proteins delivered to the apoplastic space and successful secretion are prerequisites for their functions (Wang et al., 2005). Co-expression of β-1,3-glucanase and smotin AP24 in *N. tabacum* has proven to effectively offer chloroplast-enhanced resistance against filamentous pathogens (Boccardo et al., 2019). Transgenic MuPR1 Arabidopsis enhanced resistance to *Botrytis cinerea* and *Pseudomonas syringae*, mediating oxygen radical formation and detoxification (Fang et al., 2019). Leaf proteome differentially expressed proteins, and acidic endochitinase up-regulated upon *fungal* infection and drought (Sruthilaxmi & Babu, 2020). Transcriptome analysis of poplar (Populus spp.) responses to jasmonic acid (JA) and salicylic acid (SA)/*Melampsora larici-populina* (MLP) treatments conformably regulated the concerted processes in response to fungal defence and regulated growth by integrating multiple hormone signalling pathways (Luo et al., 2019). Transcriptome analysis of *Solanum lycopersici* revealed differentially expressed genes belonging to classes including signalling, hormone response, and pathogen resistance in the presence and absence of pathogen infection (Mani et al., 2021). Transcriptome sequencing has enormously accelerated recent research on host–pathogen interactions in plants such as tomatoes (Zhang et al., 2021), potatoes (Cao et al., 2021), and rice (Cheng et al., 2020).

5.4.2 Jasmonic Acid Pathway

The jasmonic acid (3-oxo-2–20-cis-pentenyl-cyclopentane-1-acetic acid-JA) is phytohormones responding to developmental cues or environmental stresses to coordinate plant growth, development, or defence against pathogens and herbivores (Yan & Xie, 2015; Zhang et al., 2017b). Jasmonic acid and its derivatives (e.g., methyl ester, jasmonoyl-amino acid, JA-Ile methyl ester, and isoleucine conjugate) are collectively referred to as jasmonates (JAs) and play significant roles in endogenous growth-regulating substances in higher plants (Ruan et al., 2019). An amino acid-conjugated form of JA (jasmonoyl-isoleucine (JA-Ile)) stimulates plant growth and defence response to both biotic and abiotic factors (Katsir et al., 2008; Zhang et al., 2020). The jasmonic acid carboxyl methyltransferase (JMT) gene in Arabidopsis exhibits high production of methyl jasmonate (MeJA) but not JA, which constitutively expressed JA-responsive genes variant-specific surface proteins (VSP) and plant defence in 1.2 (PDF1.2) (Seo et al., 2001). Moreover, those transgenic plants could defend themselves against infection by *Botrytis cinerea*, indicating that this methylated form of JA is one of the active molecules in plant defence response (Seo et al., 2001). Several transcription factors (TFs) responsible for the activation of different JA-mediated responses have been identified. Among the TFs family, six prominent TF families – APETALA2/ethylene responsive factor (AP2/ERF) family transcription factors (AP2/ERFs), bHLH (basic helix-loop-helix), MYB (myeloblastosis-related), NAC (no apical meristem), Arabidopsis transcription activation factor (ATAF1/2), and cup-shaped cotyledon (CUC2), WRKY, and bZIP (basic leucine zipper) – are involved in abiotic and biotic stress responses (Shao et al., 2015; Ng et al., 2018).

Transcriptomic analysis of inducible defence genes in response to MeJA and Asian corn borer attack in *Zea mays* found that many essential defence genes were up-regulated between stalk borer feeding and exogenous application of MeJA (Yang et al., 2015). Two WRKY transcription factors were also up-regulated in the

JA-signalling pathway in *Arabidopsis* (Journot-Catalino et al., 2006). Proteomic changes in *Zea mays* in response to MeJA treatment were found in 62 MeJA-responsive proteins, and among 43 proteins were increased expression-level changes. Similarly, the same overexpressed proteins inhibit *Ostrinia furnacalis*, the Asian corn borer (ACB) (Zhang et al., 2015).

5.5 PLANT RESPONSE TO STRESS (INDUCED DEFENCE MECHANISMS)

5.5.1 Plant Stress Detection

Detection of plant stress due to pathogen infection is vital for improving crop yield in the current scenario (Galieni et al., 2021). The early detection of pathogens is essential in plant disease management and minimizing the spreading of new diseases. It can be done using both direct and indirect methods. Disease-causing pathogens such as bacteria, fungi, and viruses are directly detected to identify the disease/pathogen accurately. Alternatively, indirect methods identify through various parameters, such as changes in the biochemical profiles and volatile organic compounds released by infected plants. Several accurate, rapid, automated techniques were used to detect the most affecting pathogens in crops at the various plant stages, ranging from qualitative non-destructive remote sensing techniques (Vis/NIR spectroscopy, fluorescence spectroscopy, fluorescence imaging, multi- and hyperspectral imaging, and thermal imaging) to sample destructive quantitative techniques (molecular methods and metabolomics), to frontier quantitative and non-disruptive functional imaging technologies (Galieni et al., 2021). In recent years, several molecular tools have been developed to detect plant stress caused by pathogens. The most commonly used molecular methods, such as polymerase chain reaction (PCR and RT-qPCR) and enzyme-linked immunosorbent assay (ELISA), are mainly applied for the early diagnosis of plant diseases caused by several pathogens, including immunofluorescence imaging (Pérez-Bueno et al., 2016; Gautam et al., 2020), fluorescence in situ hybridization (FISH; Bellgard et al., 2016), flow cytometry method based on laser technique (Chitarra & van den Bulk, 2003), and DNA microarrays.

Among molecular techniques, omics technologies are effective diagnostic tools for plant pathogen detection. A comprehensive understanding of plant health and diseases requires the interpretation of molecular intricacy and variations at multiple levels, such as genome, transcriptome, proteome, and metabolome. Integrated approaches combine individual omics data to understand the interplay of molecules sequentially or simultaneously.

5.5.1.1 Plant Stress Detection: Metabolomics

Metabolomics studies identify changes in primary or specialized metabolism to the defence responses of plants against bacterial, fungal, and viral pathogens (Castro-Moretti et al., 2020). Plants depend on specific survival strategies against stress, and the defence response results in the production of various primary and secondary metabolites (Stagnari et al., 2016). Primary metabolites such as carbohydrates and amino acids combat plant stress (Watanabe et al., 2014). The sugar levels of infected plant tissues activate plant defence responses, which regulate hormonal signalling and induce plant resistance against pathogens (Morkunas & Ratajczak, 2014). The critical metabolic pathways, such as carbohydrate metabolism, amino acid metabolism, galactose metabolism, glyoxylate and dicarboxylate metabolism, starch and sucrose metabolism, ABC transporters, etc., are activated against pathogens (Morkunas & Ratajczak, 2014). Therefore, understanding related pathways in carbohydrate metabolism, amino acid metabolism, and ABC transporters could provide insight into a specific genetic modification platform to strengthen plant resistance mechanisms. Advanced metabolomics analytical platforms, nuclear magnetic resonance spectroscopy (NMR), mass spectroscopy (MS), direct flow injection (DFI) mass spectrometry, high-resolution mass spectroscopy (HRMS), direct infusion mass spectroscopy (DIMS), Fourier-transform infrared spectroscopy (FTIR), high-performance thin-layer chromatography (HPTLC), Fourier-transform ion cyclotron resonance mass spectroscopy (FI-ICR-MS), ultra-performance liquid chromatography (UPLC), capillary electrophoresis mass spectroscopy (CE-MS), gas chromatography mass spectroscopy (GC-MS), and liquid chromatography mass spectroscopy (LC-MS) are the most-explored analytical platform for deciphering the metabolic changes to pathogenic stresses (Piasecka et al., 2019). GC and NMR aid in elucidating the structure of metabolites and the biomolecular composition of plant extracts (Roessner et al., 2001; Fernie et al., 2004). FTIR is used to determine the infrared absorption frequency and intensity to determine the functional groups of the components in the sample and the vibration of the high polarity bonds (Chen et al., 2019).

Several metabolites have been recognized as metabolic biomarkers, including volatile organic compounds (VOCs) such as isoprene, methanol, phytohormone ethylene, and some monoterpenes, methyl jasmonate, methyl salicylate (Ninkovic et al., 2019; Castro-Moretti et al., 2020). In bacterially infected tomatoes, the level of amino acids, organic acids, rutin, and phenylpropanoids increased, while viroid infection affected only glucose and malic acid biosynthesis (López-Gresa et al., 2010). In rice, 16 fatty acids and two amino acids (glutamine and phenylalanine) were identified as resistance markers (Agarrwal et al., 2014). Non-polymerized lignin precursors in barley, rice, and purple false brome grass activate the defence-related pathways against *Magnaporthe oryzae* (Parker et al., 2009). In addition, phytoalexins and phytoanticipins are biosynthesized in response or advance to pathogen perception (Schlaeppi et al., 2010). Some of the advanced metabolomic techniques employed in plant stress detection are given in Table 5.1. Up to now, metabolomics analyses of fungal pathogen–plant interactions have mainly been performed between several fungi and their host plants, especially between *Fusarium graminearum*, *Magnaporthe oryzae*, *Ustilago maydis*, *Rhizoctonia solani*, *Sclerotinia sclerotiorum*, and their hosts (Chen et al., 2019).

TABLE 5.1
Selected Recent Metabolomics Studies for Understanding Biotic Stress in Plants

Crop	Disease	Findings	Metabolomics/ Analytical Platform	References
Rice	Rice blast, *Magnaporthe oryzae*	Increased salicylic acid (SA) and jasmonic acid (JA) content enhanced interactions between JA synthesis/ signalling gene, SA synthesis/ receptor gene, raffinose/ fructose/sucrose synthase gene, and cell wall-related genes all contribute to defence response in rice	Ultra-high performance liquid chromatography-quadrupole time-of-flight mass spectrometry (UHPLC/Q-TOF-MS)	Duan et al., 2021
Tomato	Late blight, *Phytophthora infestans*	Identified metabolite tomatidine as an important biomarker of infection, saponins as early infection metabolite markers, and isocoumarin as early and late asymptomatic infection markers along the post-infection time	LC-MS and metabolic profiling by (MALDI-MS)	Galeano Garcia et al., 2018
Wheat	Fusarium head blight, *Fusarium graminearum*	Activation of cell wall fortification pathways as well as changes in amino acid metabolism, with a particular increase in the concentration of branched-chain amino acid, is observed in all wheat varieties regardless of the level of resistance to FHB	NMR	Cuperlovic-Culf et al., 2016
Common bean	White mould, *Sclerotinia sclerotiorum*	Identified an increase in bioactive jasmonate JA-Ile ((+)-7-iso-jasmonoyl-L-isoleucine), which scavenges reactive oxygen species	RNA sequencing, GC-MS	Robison et al., 2018
Soybean	Foliar blight, *Rhizoctonia solani*	Studies showed that alcohol and its associated gene product ADH (alcohol dehydrogenase) might have important roles in soybean resistance to *R. solani*	RNA-seq-1H NMR	Copley et al., 2017

5.5.1.2 Plant Stress Detection: Transcriptomics

RNA-seq studies are primarily used to reveal plant defence responses at the molecular level and to identify specific genes involved in the process. Plant pathogens have been reported to influence the primary metabolism of plants to achieve successful infection (Rojas et al., 2014). Pathogens regulate photosynthesis for several reasons, including carbon utilization, and the downregulation of photosynthesis is considered a general response of plants to susceptible pathogens (Rojas et al., 2014; Swarbrick et al., 2006).

Transcriptome profiling using next-generation sequencing technologies is employed to detect the molecular mechanisms of plant responses to pathogenic stress. Transcriptomic data generated specifically from plant–pathogen pathosystems are therefore needed to improve our understanding of the mechanisms underlying the resistance of host plants and pathogenicity of their invaders, and, finally, this will help in developing disease-resistant varieties. Transcriptomics study on pathogenic stress between resistant and susceptible varieties was carried out in crops such as wheat infected by powdery mildew, *Blumeria graminis* f. sp. tritici (Bgt) or stripe rust fungus, *Puccinia striiformis* f. sp. tritici (Pst) (Zhang et al., 2019), canola against *Sclerotinia sclerotiorum* (Chittem et al., 2020), and tobacco due to *Phytophthora nicotianae* infection (Meng et al., 2021). Transcriptomic data revealed that different genes and proteins were differentially regulated due to pathogenic stress in plants (Table 5.2). Unravelling defence response by the host on pathogen infection in tea plants in response to grey blight infection was demonstrated by transcriptomic approach (Wang et al., 2021).

TABLE 5.2
Selected Recent Transcriptomics Studies for Understanding Biotic Stress in Plants

Crop	Pathogen	Findings	Transcriptomics/ Analytical Platform	References
Sugarcane	*Sporisorium scitamineum*	NBS-LRR domain-containing proteins, Cytochrome P450, chitinase, and leucine zipper domain proteins were found	Illumina HiSeq 2000 platform	McNeil et al. (2018)
Tobacco	*Phytophthora nicotianae*	Three pathogenesis-related proteins, 11 disease-resistant proteins, 2 CNGCs, 4 RLP/RLKs, 7 calcium-dependent protein kinases, 1 protein EDS1L, 4 calcium-binding proteins, 1 mitogen-activated protein kinase, 1 mannosyltransferase, 2 WRKY transcription factors, and 1 calmodulin-like protein were involved in disease resistance	qRT-PCR	Meng et al., 2021

TABLE 5.2 *(Continued)*
Selected Recent Transcriptomics Studies for Understanding Biotic Stress in Plants

Crop	Pathogen	Findings	Transcriptomics/ Analytical Platform	References
Tobacco	Black shank, *Phytophthora nicotianae*	Found some genes encoding a chitinase	qRT-PCR and gene ontology enrichment analysis	Yang et al., 2017
Canola	*Sclerotinia sclerotiorum*	Identified the role of peroxisome-related pathways along with cell wall degradation and detoxification of host metabolites	Illumina HiSeq 2500 platform (2X 50 bp Paired-End)	Chittem et al., 2020
Wheat	Powdery mildew, *Blumeria graminis* f. sp. *Tritici* (Bgt) or stripe rust fungus, *Puccinia striiformis* f. sp. *Tritici* (Pst)	RNA-seq analysis identified more SF, SR repressor proteins, pre-mRNA-processing factors, and snRNP auxiliary factors in wheat undertaking AS events. Also detected the AS transcripts of U1, U2, and U3-related elements/proteins involved in the wheat response to fungal infection	Illumina RNA-seq	Zhang et al., 2019
Spiny-Head Mat-Rush	*Phytophthora cinnamomi*	*MAPK* 15, Callose synthase genes, 2 *PR* genes, and 5 receptor-like protein genes	qRT-PCR	Islam et al., 2018

5.5.1.3 Plant Stress Detection: Proteomics

Plant responses to pathogens depend on the genetic basis of plant–pathogen interactions, or the ability of plants to sustain themselves against pathogens with differential expressed proteins (Yang et al., 2011). After recognizing the pathogen, the host initiates a series of physiological and biochemical responses involving molecular interactions and signal transduction. Plant acclimation to stress is linked with profound changes in proteome composition (Kosová et al., 2011).

Different proteomic techniques – such as two-dimensional polyacrylamide gel electrophoresis (2D-PAGE), fluorescent two-dimensional "difference gel electrophoresis" (2D-DIGE), ICAT (isotope-coded affinity tags), ITRAQ (isobaric tagged for relative and absolute quantitation), and MudPIT (multidimensional protein identification technology) – are used for pathogen detection (Lodha et al., 2013). Recently, mass spectrometry techniques with two soft ionization techniques, namely, matrix-assisted laser desorption/ionization (MALDI) and

TABLE 5.3
Selected Recent Proteomics Studies for Understanding Biotic Stress in Plants

Crop	Disease	Findings	Proteomics/Analytical Platform	References
Rice	Sheath blight, *Rhizoctonia solani*	Significant changes in the abundance of 755 proteins primarily involved in glycine, serine, and threonine metabolism; unsaturated fatty acid biosynthesis; glyoxylate and dicarboxylate metabolism; and glycolysis/ gluconeogenesis regulation pathways	Eight-plex iTRAQ technique (isobaric tags for relative and absolute quantitation)	Ma et al., 2020
Sugarcane	Pokkah boeng disease (PBD), *Fusarium verticillioides* and *F. proliferatum*	The net photosynthetic rate (Pn), stomatal conductance (gs), and transpiration rate (E) of sugarcane plants were significantly reduced as pathogen stress increased	iTRAQ and low PH nano-HPLC-MS-MS	Shuai et al., l2020
Sugarcane	Smut, *Sporisorium scitamineum*	Twenty differential proteins associated with photosynthesis, signal transduction, and disease resistance	Two-dimensional electrophoresis (2-DE) and MALDI-TOF-TOF/MS	Que et al., 2011
Rice	Brown spot, *Cochliobolus miyabeanus*	Nineteen proteins related to metabolism, energy, photosynthesis, transcription, stress, and defences	Two-dimensional electrophoresis (2-DE) followed by identification by mass spectrometry (MALDI TOF/TOF)	Laurindo et al., 2018
Poplar forestry species	Poplar canker, *Botryosphaeria dothidea*	Antioxidants, phenylpropanoid biosynthesis, and phenylalanine metabolism carbon fixation	Nanoflow LC-MS/ MS analysis. Proteomic data analysis using Kyoto Encyclopedia of Genes and Genomes pathway analysis (KEGG)	Li et al., 2019

electron spray ionization (ESI), are mainly used for plant–pathogen interaction and stress detection (Table 5.3). The invention of time-of-flight mass spectrometry (TOF-MS) and relatively non-destructive methods to convert proteins into volatile ions caused a revolution in mass spectrometry. The protein microarray technique is used for profiling complex samples in large quantities. It provides a platform for classical and functional proteome analysis (Duburcq et al., 2004; LaBaer & Ramachandran, 2005).

ELISA and molecular hybridization can quantify viral particles or proteins based on signal intensity (Rubio et al., 2003; Rohrman et al., 2012). However, ELISA and western blotting could be used for selective protein analysis. Quantitative real-time PCR (qPCR) is a technique used for quantification of gene expression or knockdown, viral titer, with a wide range of sensitivity and other broad applications in the field of omics technology (Boonham et al., 2014; Debreczeni et al., 2011; Ferriol et al., 2011; Sharma & Dasgupta, 2012; MacKenzie et al., 2015; Herrera-Vásquez et al., 2015). Clustered regularly interspaced short palindromic repeats (CRISPR) and CRISPR-associated genes (Cas) are a prokaryotic adaptive immune system reprogrammed into a precise, simple, and efficient gene targeting technology. CRISPR/Cas-based diagnostics may be better for developing ultra-sensitive, inexpensive, and rapid non-laboratory–based detection kits for several viruses and other pathogens (Gootenberg et al., 2018; Chen et al., 2018; Myhrvold et al., 2018; Kalinina et al., 2020). Loop-mediated isothermal amplification (LAMP) is a single-tube technique that was initially developed by Notomi et al. (2000) for the amplification of DNA and as a low-cost alternative to detect several plant viruses, including potato leafroll virus (Ahmadi et al., 2013), *Citrus tristeza* virus (Selvaraj et al., 2019), potato virus X (Jeong et al., 2015), tomato torrado virus (Budziszewska et al., 2016), and rice black-streaked dwarf virus (Du et al., 2019). Nowadays, recombinase polymerase amplification (RPA), loop-mediated isothermal amplification (LAMP), and CRISPR-based tools are popular among nucleic acid-based testing (NAT) tools and may replace the costly qRT-PCR-based testing systems (Srivastava et al., 2020).

5.5.2 Plant Disease Ecology

Omics technologies have developed in recent decades to understand the role of pathogens/endophytes in plant disease. It is possible to have a broad overview of the linkage between genotypes and phenotypes to understand protection and health (Guillemin et al., 2016). Plant pathogens are parasites that live in a host. Most plant pathogen species can attack a broad diversity of plant species and play a vital role in the structure, dynamics, and evolution of natural plant communities concerning global changes (Gilbert, 2002; Crandall et al., 2020). Plant pathogens are responsible for numerical changes in host populations and evolutionary changes through selection for resistant genotypes.

Similarly, plant rhizosphere and phyllosphere microbes are inhabited internally by endophytes. Endophytes are abundant in plants, and studies continuously emanate their ability to protect plants from pathogens (Fadiji & Babalola, 2020). Most endophytes act as commensals without any known effect on their plant host. Still, multiple bacteria and fungi establish a mutualistic relationship with plants, and some act as pathogens (Brader et al., 2017). Most of the research focuses on the function of one or a handful of endophytes rather than entire communities. Endophytic *Burkholderia phytofirmans* PsJN increased metabolic activity inside plants, changed plant physiology due to plant stress, and adjusted their gene expression pattern to the altered conditions. A guiding principle is important in understanding the spatial and temporal scales at which plants and pathogens interact (Sheibani-Tezerji et al., 2015; Chitnis et al., 2020). The term omics, which looks into the global profiling and analysis of various plant–pathogen interactions, evolution, and ecology, has gained new heights to advance next-generation sequencing. It has broader application in the genetic improvement of crops for resistance to various diseases as well as economic significance (Bhadauria, 2016), using evolutionary models, under different constraint levels, mirroring the evolution of the underlying traits that required disease resistance. Utilization of transcriptomic and metabolomics (pathogens)/spectranomics (plants) can provide novel insights into plant–pathogen evolution and diversity. Omics could be applied to plant disease ecology to understand the evolutionary dynamics of beneficial plant–microbe relationships.

5.6 CHALLENGES

Omics approaches help to understand the interaction of pathogens with the environment and plants and their functions. However, while designing projects using omics technologies, there is a need to switch from observational to hypothesis-driven research. These approaches depend on the highest computational tools and well-equipped laboratory facilities to understand plant stress responses and the functional changes occurring in host pathogens due to their interaction. Even though progress has been made in proteomic analysis of multi-trophic interaction of microbial species, there are technical challenges in detecting the same type of proteins as protein separation and analysis of results are skill-based and challenging to automate (Massart et al., 2015).

5.7 CONCLUSION AND FUTURE PERSPECTIVES

Omics-based studies facilitate a greater understanding of plant–pathogen interactions and have become the mainstream approach to rapidly identifying candidate/pathogenicity genes in the metabolic pathway overviews of the transcriptional changes. Bioinformatic methods will promote transcriptional changes in response to infection and provide helpful information for predicting transcriptional factors involved in defence responses. This genomics/proteomics/metabolomics "revolution" made an essential contribution to plant pathology, rapidly increasing our knowledge of the molecular mechanisms underpinning pathogenesis, resistance, and the mode of action of beneficial microorganisms. Omics technologies will describe the background of plant–pathogen interaction, are essential for understanding the molecular mechanism of the development of plant diseases, and will provide helpful information for guiding the development of disease-resistant varieties. Future perspectives include the development of omics mega-libraries to determine and interpret plant–pathogen interaction. Further studies should ultimately contribute to providing databases that allow functional comparisons and integration of plant–pathogen interaction to improve productivity and adaptability of crops to environmental stress. Omics technology must strengthen the research groups dedicated to studying plant–pathogen interaction.

ACKNOWLEDGEMENT

SG, AC, and AR would like to acknowledge "EVA4.0", No. CZ.02.1.01/0.0/0.0/16_019/0000803, financed by "OP RDE" and "EXTEMIT – K", No. CZ.02.1.01/0.0/0.0/15_003/0000433, financed by OP RDE and Excellent Team Grants (2023-2024), Faculty of Forestry and Wood Sciences, Czech University of Life Sciences, Prague, for financial support during the preparation of this book chapter.

ABBREVIATIONS

2-DE	two-dimensional gel electrophoresis
CAMP	Collection of Anti-Microbial Peptides
CHS/CHI	chalcone synthase/chalcone isomerase
DArT	diversity arrays technology
DGGE	denaturing gradient gel electrophoresis
DIGE	difference gel electrophoresis
ESI	electrospray ionization
FISH	fluorescent in situ hybridization
GCMS	gas chromatography and mass spectrometry
GC-MS	gas chromatography and mass spectrometry
GWAS	genome-wide association mapping
H-NMR	spectroscopy – nuclear magnetic resonance spectroscopy
IT-MS	ion trap mass spectrometry
ITS	intergenic transcribed spacers
LC-MS/MS	liquid chromatography mass spectrometry
LIDAR	light detection and ranging
MABC	marker-assisted breeding (MBC), marker-assisted backcross breeding
MALDI-MS	matrix-assisted laser desorption/ionization mass spectrometry
MALDI-TOF-TOF/MS	tandem time-of-flight mass spectrometry
NMR	nuclear magnetic resonance
PacBio	Pacific Biosciences

PGM	personal genome machine
PHI-base	pathogen–host interaction database
qTOF HD-MS	quadrupole time-of-flight high-definition mass spectrometer
RAPD	random amplified polymorphic DNA
RFLP	restriction fragment length polymorphism
SAGE	serial analysis of gene expression
SAMPL	selective amplification of microsatellite polymorphic loci
SRAP	sequence-related amplified polymorphism
TRAP	target region amplification polymorphism (TRAP)

REFERENCES

Agarrwal, R., Bentur, J. S., & Nair, S. (2014, September). Gas chromatography mass spectrometry based metabolic profiling reveals biomarkers involved in rice-gall midge interactions. *Journal of Integrative Plant Biology*, *56*(9), 837–848. https://doi.org/10.1111/jipb.12244

Ahmadi, S., Almasi, M. A., Fatehi, F., Struik, P. C., & Moradi, A. (2013). Visual detection of potato leafroll virus by one-step reverse transcription loop-mediated isothermal amplification of DNA with hydroxynaphthol blue dye. *Journal of Phytopathology*, *161*, 120–124. https://doi.org/10.1111/jph.12037

Altelaar, A. F., Munoz, J., & Heck, A. J. (2013, January). Next-generation proteomics: Towards an integrative view of proteome dynamics. *Nature Reviews Genetics*, *14*(1), 35–48. https://doi.org/10.1038/nrg3356

Araus, J. L., & Cairns, J. E. (2014, January). Field high-throughput phenotyping: The new crop breeding frontier. *Trends in Plant Science*, *19*(1), 52–61. https://doi.org/10.1016/j.tplants.2013.09.008

Arentz, G., Weiland, F., Oehler, M. K., & Hoffmann, P. (2015, April). State of the art of 2D DIGE. *Proteomics Clinical Applications*, *9*(3–4), 277–288. https://doi.org/10.1002/prca.201400119

Assche, R. V., Broeckx, V., Boonen, K., Maes, E., De Haes, W., Schoofs, L., & Temmerman, L. (2015). Integrating -omics: Systems biology as explored through C. elegans research. *Journal of Molecular Biology*, *23*(21), 427, 3441–3451. https://doi.org/10.1016/j.jmb.2015.03.015

Audic, S., & Claverie, J. M. (1997, October). The significance of digital gene expression profiles. *Genome Research*, *7*(10), 986–995. https://doi.org/10.1101/gr.7.10.986. PMID: 9331369.

Badjakov, I., Kondakova, V., & Atanassov, A. (2012). Metabolomics: Current view on fruit quality in relation to human health. In N. Benkeblia (Ed.), *Sustainable agriculture and new biotechnologies* (pp. 303–319). CRC Press.

Bainbridge, M. N., Warren, R. L., Hirst, M., Romanuik, T., Zeng, T., Go, A., Delaney, A., Griffith, M., Hickenbotham, M., Magrini, V., & Mardis, E. R. (2006, September 29). Analysis of the prostate cancer cell line LNCaP transcriptome using a sequencing-by-synthesis approach. *BMC Genomics*, *7*, 246. https://doi.org/10.1186/1471-2164-7-246

Barbulovic-Nad, I., Lucente, M., Sun, Y., Zhang, M., Wheeler, A. R., & Bussmann, M. (2006, October–December). Bio-microarray fabrication techniques – a review. *Critical Reviews in Biotechnology*, *26*(4), 237–259. https://doi.org/10.1080/07388550600978358

Baxter, A., Mittler, R., & Suzuki, N. (2014, March). ROS as key players in plant stress signalling. *Journal of Experimental Botany*, *65*(5), 1229–1240. https://doi.org/10.1093/jxb/ert375

Behnke, K., Kaiser, A., Zimmer, I., Brüggemann, N., Janz, D., Polle, A., Hampp, R., Hänsch, R., Popko, J., Schmitt-Kopplin, P., & Ehlting, B. (2010, September). RNAi-mediated suppression of isoprene emission in poplar transiently impacts phenolic metabolism under high temperature and high light intensities: A transcriptomic and metabolomic analysis. *Plant Molecular Biology*, *74*(1–2), 61–75. https://doi.org/10.1007/s11103-010-9654-z

Bellgard, S. E., Padamsee, M., Probst, C., Lebel, T., & Williams, S. (2016). Visualizing the early infection of *Agathis australis* by *Phytophthora agathidicida*, using microscopy and fluorescent in situ hybridization. *Forest Pathology*, *46*, 622–631.

Berens, A. E., Jensen, S. K. G., & Nelson III, C. A. (2017, July 20). Biological embedding of childhood adversity: From physiological mechanisms to clinical implications. *BMC Medicine*, *15*(1), 135. https://doi.org/10.1186/s12916-017-0895-4

Bhadauria, V. (2016). OMICS in plant disease resistance. *Current Issues in Molecular Biology*, *19*, 1–2. Epub September 11, 2015.

Boccardo, N. A., Segretin, M. E., Hernandez, I., Mirkin, F. G., Chacon, O., Lopez, Y, BorrasHidalgo, O., & Bravo-Almonacid, F. F. (2019). Expression of pathogenesis-related proteins in transplastomic tobacco plants confers resistance to filamentous pathogens under field trials. *Scientific Reports*, *9*, 2791.

Bollina, V., Kumaraswamy, G. K., Kushalappa, A. C., Choo, T. M., Dion, Y., Rioux, S., Faubert, D., & Hamzehzarghani, H. (2010, November). Mass spectrometry-based metabolomics application to identify quantitative resistance-related metabolites in barley against Fusarium head blight. *Molecular Plant Pathology*, *11*(6), 769–782. https://doi.org/10.1111/j.1364-3703.2010.00643.x

Boonham, N., Kreuze, J., Winter, S., van der Vlugt, R., Bergervoet, J., Tomlinson, J., & Mumford, R. (2014, June 24). Methods in virus diagnostics: From ELISA to next generation sequencing. *Virus Research*, *186*, 20–31. https://doi.org/10.1016/j.virusres.2013.12.007

Bostock, R. M. (2005). Signal crosstalk and induced resistance: Straddling the line between cost and benefit. *Annual Review of Phytopathology*, *43*, 545–580. https://doi.org/10.1146/annurev.phyto.41.052002.095505

Boyd, L. A., Ridout, C., O'Sullivan, D. M., Leach, J. E., & Leung, H. (2013). Plant-pathogen interactions: Disease resistance in modern agriculture. *Trends in Genetics*, *29*(4), 233–240. https://doi.org/10.1016/j.tig.2012.10.011

Brader, G., Compant, S., Vescio, K., Mitter, B., Trognitz, F., Ma, L. J., & Sessitsch, A. (2017, August 4). Ecology and genomic insights into plant-pathogenic and plant-nonpathogenic endophytes. *Annual Review of Phytopathology*, *55*, 61–83. https://doi.org/10.1146/annurev-phyto-080516-035641

Brichet, N., Fournier, C., Turc, O., Strauss, O., Artzet, S., Pradal, C., Welcker, C., Tardieu, F., & Cabrera-Bosquet, L. (2017). A robot-assisted imaging pipeline for tracking the growths of maize ear and silks in a high-throughput phenotyping platform. *Plant Methods*, *13*, 96. https://doi.org/10.1186/s13007-017-0246-7

Broughton, W. J., Hernandez, G., Blair, M., Stephen, B., & Vanderleyden, J. (2003). Beans (*Phaseolus* spp.): Model food legumes. *Plant and Soil*, *252*, 74.

Budziszewska, M., Wieczorek, P., & Obrępalska-Stęplowska, A. (2016). One-step reverse transcription loop-mediated isothermal amplification (RT-LAMP) for detection of tomato torrado virus. *Archives of Virology*, *161*, 1359–1364. https://doi.org/10.1007/s00705-016-2774-2

Cao, W., Gan, L., Wang, C., Zhao, X., Zhang, M., Du, J., Zhou, S., & Zhu, C. (2021, February 10). Genome-wide identification and characterization of potato long non-coding RNAs associated with phytophthora infestans resistance. *Frontiers in Plant Science, 12*, 619062. https://doi.org/10.3389/fpls.2021.619062

Castro-Moretti, F. R., Gentzel, I. N., Mackey, D., & Alonso, A. P. (2020, January 29). Metabolomics as an emerging tool for the study of plant-pathogen interactions. *Metabolites, 10*(2), 52. https://doi.org/10.3390/metabo10020052

Catola, S., Marino, G., Emiliani, G., Huseynova, T., Musayev, M., Akparov, Z., & Maserti, B. E. (2016, February). Physiological and metabolomic analysis of *Punica granatum* (L.) under drought stress. *Planta, 243*(2), 441–449. https://doi.org/10.1007/s00425-015-2414-1

Chen, F., Ma, R., & Chen, X. L. (2019, September 4). Advances of metabolomics in fungal pathogen-plant interactions [published correction appears in *Metabolites, 9*(9)]. *Metabolites, 9*(8), 169. Published August 15, 2019. https://doi.org/10.3390/metabo9080169

Chen, J. S., Ma, E., Harrington, L. B., Da Costa, A, Tian, X., Palesfsky, J. M., & Doudna, J. A. (2018, April 27). CRISPR-Cas12a target binding unleashes indiscriminate single-stranded dnase activity. *Science, 360*, 436–439. https://doi.org/10.1126/science.aar6245

Cheng, A. P., Chen, S. Y., Lai, M. H., Wu, D. H., Lin, S. S., Chen, C. Y., & Chung, C. L. (2020). Transcriptome analysis of early defenses in rice against *Fusarium fujikuroi*. *Rice, 13*, 65. https://doi.org/10.1186/s12284-020-00426-z

Chisholm, S. T., Coaker, G., Day, B., & Staskawicz, B. J. (2006, February 24). Host-microbe interactions: Shaping the evolution of the plant immune response. *Cell, 124*(4), 803–814. https://doi.org/10.1016/j.cell.2006.02.008

Chitarra, L. G., & van den Bulk, R. W. (2003). The application of flow cytometry and fluorescent probe technology for detection and assessment of viability of plant pathogenic bacteria. *European Journal of Plant Pathology, 109*, 407–417. https://doi.org/10.1023/A:1024275610233

Chitnis, V. R., Suryanarayanan, T. S., Nataraja, K. N., Prasad, S. R., Oelmüller, R., & Shaanker, R. U. (2020, October 27). Fungal endophyte-mediated crop improvement: The way ahead. *Frontiers in Plant Science, 11*, 561007. https://doi.org/10.3389/fpls.2020.561007

Chittem, K., Yajima, W. R., Goswami, R. S., & Del Río Mendoza, L. E. (2020, March 11). Transcriptome analysis of the plant pathogen *Sclerotinia sclerotiorum* interaction with resistant and susceptible canola (Brassica napus) lines. *PLoS One, 15*(3), e0229844. https://doi.org/10.1371/journal.pone.0229844

Cohen, S. P., & Leach, J. E. (2019). Abiotic and biotic stresses induce a core transcriptome response in rice. *Scientific Reports, 9*, 6273. https://doi.org/10.1038/s41598-019-42731-8

Coleman-Derr, D., & Tringe, S. G. (2014, June 6). Building the crops of tomorrow: Advantages of symbiont-based approaches to improving abiotic stress tolerance. *Frontiers in Microbiology, 6*(5), 283. https://doi.org/10.3389/fmicb.2014.00283

Colwell, R. R., Brayton, P., Herrington, D., Tall, B., Huq, A., & Levine, M. M. (1996). Viable but non-culturable Vibrio cholerae O1 revert to a cultivable state in the human intestine. *World Journal of Microbiology and Biotechnology, 12*(1), 28–31. https://doi.org/10.1007/BF00327795

Confalonieri, R., Paleari, L., Foi, M., Movedi, E., Vesely, F. M., Thoelke W., Agape, C., Borlini, G., Ferri, I., Massara, F., & Motta, R. (2017). Pocketplant3d: Analysing canopy structure using a smartphone. *Biosystems Engineering, 164*, 1–12. https://doi.org/10.1016/j.biosystemseng.2017.09.014

Cook, D., Fowler, S., Fiehn, O., & Thomashow, M. F. (2004). A prominent role for the CBF cold responsive pathway in configuring the low-temperature metabolome of Arabidopsis. *Proceedings of the National Academy of Sciences of the United States of America*, *101*, 15243–15248.

Copley, T. R., Aliferis, K. A., Kliebenstein, D. J., & Jabaji, S. H. (2017, April 27). An integrated RNAseq-1H NMR metabolomics approach to understand soybean primary metabolism regulation in response to *Rhizoctonia* foliar blight disease. *BMC Plant Biology*, *17*(1), 84. https://doi.org/10.1186/s12870-017-1020-8

Crandall, S. G., Gold, K. M., Jiménez-Gasco, M. D. M., Filgueiras, C. C., & Willett, D. S. (2020). A multi-omics approach to solving problems in plant disease ecology. *PLoS One*, *15*(9), e0237975. Published September 22, 2020. https://doi.org/10.1371/journal.pone.0237975

Cuperlovic-Culf, M., Wang, L., Forseille, L., Boyle, K., Merkley, N., Burton, I., & Fobert, P. R. (2016). Metabolic biomarker panels of response to fusarium head blight infection in different wheat varieties. *PLoS One*, *11*(4), e0153642. Published April 21, 2016. https://doi.org/10.1371/journal.pone.0153642

Debreczeni, D. E., Ruiz-Ruiz, S., Aramburu, J., López, C., Belliure, B., Galipienso, L., Soler, S., & Rubio, L. (2011, September). Detection, discrimination and absolute quantitation of Tomato spotted wilt virus isolates using real time RT-PCR with TaqMan(®)MGB probes. *Journal of Virological Methods*, *176*(1–2), 32–37. https://doi.org/10.1016/j.jviromet.2011.05.027

Dempsey, D. A., & Klessig, D. F. (2017, March 23). How does the multifaceted plant hormone salicylic acid combat disease in plants and are similar mechanisms utilized in humans? *BMC Biology*, *15*(1), 23. https://doi.org/10.1186/s12915-017-0364-8

Dempsey, D. A., Vlot, A. C., Wildermuth, M. C., & Klessig, D. F. (2011). Salicylic acid biosynthesis and metabolism. *Arabidopsis Book*, *9*, e0156. https://doi.org/10.1199/tab.0156

De Vleesschauwer, D., Xu, J., & Höfte, M. (2014). Making sense of hormone-mediated defense networking: From rice to Arabidopsis. *Frontiers in Plant Science*, *5*, 611. Published November 11, 2014. https://doi.org/10.3389/fpls.2014.00611

Du, L., Shi, W., Li, X., Lan, Y., Sun, F., Fan, Y., Zhou, T., & Zhou, Y. (2019). A reverse-transcription loop-mediated isothermal amplification (RT-LAMP) assay for detecting the pathogen of maize rough dwarf disease in China. *Australasian Plant Pathology*, *48*, 485–489. https://doi.org/10.1007/s13313-019-00649-9

Duan, G., Li, C., Liu, Y., Ma, X., Luo, Q., & Yang, J. (2021, January 11). Magnaporthe oryzae systemic defense trigger 1 (MoSDT1)-mediated metabolites regulate defense response in rice. *BMC Plant Biology*, *21*(1), 40.

Duburcq, X., Olivier, C., Malingue, F., Desmet, R., Bouzidi, A., Zhou, F., Auriault, C., & Gras-Masse, H. (2004). Peptide-protein microarrays for the simultaneous detection of pathogen infections. *Bioconjugate Chemistry*, *15*(2), 307–316. doi:10.1021/bc034226d

Duque, A. S., de Almeida, A. M., da Silva, A. B., da Silva, J. M., Farinha, D. A. P., Santos, P. F., & Araujo, S. S. (2013). Abiotic stress responses in plants: unraveling the complexity of genes and networks to survive. *Abiotic Stress-Plant Responses and Applications in Agriculture*, 49–101.

Fadiji, A. E., & Babalola, O. O. (2020, December). Exploring the potentialities of beneficial endophytes for improved plant growth. *Saudi Journal of Biological Sciences*, *27*(12), 3622–3633. https://doi.org/10.1016/j.sjbs.2020.08.002

Fang, L. J., Qin, R. L., Liu, Z., Liu, C. R., Gai, Y. P., & Ji, X. L. (2019). Expression and functional analysis of a PR-1 gene, MuPR1, involved in disease resistance response in mulberry (*Morus multicaulis*). *Journal of Plant Interactions*, *14*, 376–385.

Fernie, A. R., Trethewey, R. N., Krotzky, A. J., & Willmitzer, L. (2004, September). Metabolite profiling: From diagnostics to systems biology. *Nature Reviews Molecular Cell Biology*, *5*(9), 763–769. https://doi.org/10.1038/nrm1451

Ferriol, I., Ruiz-Ruiz, S., & Rubio, L. (2011, November). Detection and absolute quantitation of Broad bean wilt virus 1 (BBWV-1) and BBWV-2 by real time RT-PCR. *Journal of Virological Methods*, 177(2), 202–205. https://doi.org/10.1016/j.jviromet.2011.08.003

Fichot, E. B., & Norman, R. S. (2013, March 4). Microbial phylogenetic profiling with the Pacific Biosciences sequencing platform. *Microbiome*, *1*(1), 10. https://doi.org/10.1186/2049-2618-1-10

Fiorani, F., & Schurr, U. (2013). Future scenarios for plant phenotyping. *Annual Review of Plant Biology*, *64*, 267–291. https://doi.org/10.1146/annurev-arplant-050312-120137

Fiorilli, V., Vannini, C., Ortolani, F., Garcia-Seco, D., Chiapello, M., Novero, M., Domingo, G., Terzi, V., Morcia, C., Bagnaresi, P., Moulin, L., Bracale, M., & Bonfante, P. (2018, June 25). Omics approaches revealed how arbuscular mycorrhizal symbiosis enhances yield and resistance to leaf pathogen in wheat. *Scientific Reports*, *8*(1), 9625. https://doi.org/10.1038/s41598-018-27622-8

Fritsche-Neto, R., & Borém, A. (2015). *Phenomics how next-generation phenotyping is revolutionizing plant breeding*. Springer.

Galeano Garcia, P., Neves Dos Santos, F., Zanotta, S., Eberlin, M. N., & Carazzone, C. (2018, December 15). Metabolomics of solanum lycopersicum infected with phytophthora infestans leads to early detection of late blight in asymptomatic plants. *Molecules*, *23*(12), 3330. https://doi.org/10.3390/molecules23123330

Galieni, A., D'Ascenzo, N., Stagnari, F., Pagnani, G., Xie, Q., & Pisante, M. (2021, January 27). Past and future of plant stress detection: An overview from remote sensing to positron emission tomography. *Frontiers in Plant Science*, *11*, 609155. https://doi.org/10.3389/fpls.2020.609155

Gao, Q. M., Niu, J., Zhao, S., Jiao, C., Xu, W., Fei, Z., & Wang, X. (2012). Characterization of Erysiphe necator-responsive genes in Chinese Wild Vitis quinquangularis. *International Journal of Molecular Sciences*, *13*(9), 11497–11519. https://doi.org/10.3390/ijms130911497

Gao, Q. M., Zhu, S., Kachroo, P., & Kachroo, A. (2015, April 13). Signal regulators of systemic acquired resistance. *Frontiers in Plant Science*, *6*, 228. https://doi.org/10.3389/fpls.2015.00228

Gautam, R., Parajuli, K., Tshokey, T., John, S., & Jeevan, S. (2020). Diagnostic evaluation of IgM ELISA and IgM Immunofluorescence assay for the diagnosis of Acute Scrub Typhus in central Nepal. *BMC Infectious Diseases*, 20, 138. https://doi.org/10.1186/s12879-020-4861-y

Gilbert, G. S. (2002). Evolutionary ecology of plant diseases in natural ecosystems. *Annual Review of Phytopathology*, *40*, 13–43. https://doi.org/10.1146/annurev.phyto.40.021202.110417

Gilbert, G. S., & Parker, I. M. (2010, March). Rapid evolution in a plant-pathogen interaction and the consequences for introduced host species. *Evolutionary Applications*, *3*(2), 144–156. https://doi.org/10.1111/j.1752-4571.2009.00107.x

Glazebrook, J. (2005). Contrasting mechanisms of defense against biotrophic and necrotrophic pathogens. *Annual Review of Phytopathology*, *43*, 205–227. https://doi.org/10.1146/annurev.phyto.43.040204.135923

Glenn, T. (2014). 2014 NGS field guide: Overview. *The Molecular Ecologist*. www.molecularecologist.com/next-gen-fieldguide-2014

Göhre, V., & Robatzek, S. (2008). Breaking the barriers: Microbial effector molecules subvert plant immunity. *Annual Review of Phytopathology*, *46*, 189–215. https://doi.org/10.1146/annurev.phyto.46.120407.110050

Gootenberg, S., Abudayyeh, O. O., Kellner, M. J., Joung, J., Collins, J. J., & Zhang, F. (2018, April 27). Multiplexed and portable nucleic acid detection platform with Cas13, Cas12a, and Csm6. *Science*, *360*, 439–444. https://doi.org/10.1126/science.aaq0179

Guillemin, N., Horvatić, A., Kuleš, J., Galan, A., Mrljak, V., & Bhide, M. (2016, June 21). Omics approaches to probe markers of disease resistance in animal sciences. *Molecular Omics*, *12*(7), 2036–2346. https://doi.org/10.1039/c6mb00220j

Hammami, R., Ben Hamida, J., Vergoten, G., & Fliss, I. (2009, January). PhytAMP: A database dedicated to antimicrobial plant peptides. *Nucleic Acids Research*, *37*(Database issue), D963–D968. https://doi.org/10.1093/nar/gkn655. Epub October 4, 2008.

Han, X., & Kahmann, R. (2019, June 26). Manipulation of phytohormone pathways by effectors of filamentous plant pathogens. *Frontiers in Plant Science*, *10*, 822. https://doi.org/10.3389/fpls.2019.00822

Handelsman, J., Rondon, M. R., Brady, S. F., Clardy, J., & Goodman, R. M. (1998). Molecular biological access to the chemistry of unknown soil microbes: A new frontier for natural products. *Chemical Biology*, *5*(10), R245–R249. https://doi.org/10.1016/s1074-5521(98)90108-9

Heather, J. M., & Chain, B. (2016, January). The sequence of sequencers: The history of sequencing DNA. *Genomics*, 107(1), 1–8. https://doi.org/10.1016/j.ygeno.2015.11.003

Helmy, M., Tomita, M., & Ishihama, Y. (2011). OryzaPG-DB: Rice proteome database based on shotgun proteogenomics. *BMC Plant Biology*, *11*, 63.

Herrera-Vásquez, J. A., Rubio, L., Alfaro-Fernández, A., Debreczeni, D. E., Font-San-Ambrosio, I., Falk, B. W., & Ferriol, I. (2015, September 1). Detection and absolute quantitation of Tomato torrado virus (ToTV) by real time RT-PCR. *Journal of Virological Methods*, 221, 90–94. https://doi.org/10.1016/j.jviromet.2015.04.029

Houle, D., Govindaraju, D. R., & Omholt, S. (2010, December). Phenomics: The next challenge. *Nature Reviews Genetics*, *11*(12), 855–866. https://doi.org/10.1038/nrg2897

Islam, M. T., Hussain, H. I., Rookes, J. E., & Cahill, D. M. (2018, January). Transcriptome analysis, using RNA-Seq of Lomandra longifolia roots infected with Phytophthora cinnamomi reveals the complexity of the resistance response. *Plant Biology (Stuttg)*, *20*(1), 130–142. https://doi.org/10.1111/plb.12624

Islam, Z. F., Cordero, P. R. F., & Greening, C. (2019, November 22). Putative iron-sulfur proteins are required for hydrogen consumption and enhance survival of mycobacteria. *Frontiers in Microbiology*, *10*, 2749. https://doi.org/10.3389/fmicb.2019.02749

Jannink, J. L., Lorenz, A. J., & Iwata, H. (2010, March). Genomic selection in plant breeding: From theory to practice. *Briefings in Functional Genomics*, *9*(2), 166–177. https://doi.org/10.1093/bfgp/elq001

Jeong, J., Cho, S. Y., Lee, W. H., Lee, K. J., & Ju, H. J. (2015, September). Development of a rapid detection method for potato virus X by reverse transcription loop-mediated isothermal amplification. *The Plant Pathology Journal*, *31*(3), 219–225. https://doi.org/10.5423/PPJ.OA.03.2015.0044

Jogaiah, S., Govind, S. R., & Tran, L. S. (2013, March). Systems biology-based approaches toward understanding drought tolerance in food crops. *Critical Reviews in Biotechnology*, *33*(1), 23–39. https://doi.org/10.3109/07388551.2012.659174

Jones, J. D., & Dangl, J. L. (2006, November 16). The plant immune system. *Nature*, *444*(7117), 323–329. https://doi.org/10.1038/nature05286

Journot-Catalino, N., Somssich, I. E., Roby, D., & Kroj, T. (2006, November). The transcription factors WRKY11 and WRKY17 act as negative regulators of basal resistance in *Arabidopsis thaliana*. *Plant Cell*, *18*(11), 3289–3302. https://doi.org/10.1105/tpc.106.044149

Kalinina, N. O., Khromov, A., Love, A. J., Taliansky, M. E. (2020). CRISPR applications in plant virology: Virus resistance and beyond. *Phytopathology*, *110*, 18–28. https://doi.org/10.1094/PHYTO-07-19-0267-IA

Kang, W. H., Sim, Y. M., Koo, N., Nam, J. Y., Lee, J., Kim, N., Jang, H., Kim, Y., & Yeom, S. (2020). Transcriptome profiling of abiotic responses to heat, cold, salt, and osmotic stress of *Capsicum annuum* L. *Scientific Data*, *7*, 17.

Katsir, L., Chung, H. S., Koo, A. J., & Howe, G. A. (2008, August). Jasmonate signaling: A conserved mechanism of hormone sensing. *Current Opinion in Plant Biology*, *11*(4), 428–435. https://doi.org/10.1016/j.pbi.2008.05.004

Kaya, H. B., Cetin, O., Kaya, H., Sahin, M., Sefer, F., Kahraman, A., & Tanyolac, B. (2013). SNP discovery by Illumina-based transcriptome sequencing of the olive and the genetic characterization of Turkish olive genotypes revealed by AFLP, SSR and SNP markers. *PLoS One*, *8*, e73674.

Kazan, K. (2015). Diverse roles of jasmonates and ethylene in abiotic stress tolerance. *Trends in Plant Science*, *20*(4), 219–229. https://doi.org/10.1016/j.tplants.2015.02.001

Keller, E. F. (2011). Genes, genomes, and genomics. *Biological Theory*, *6*, 132–140. https://doi.org/10.1007/s13752-012-0014-x

Klukas, C., Chen, D., & Pape, J. M. (2014, June). Integrated analysis platform: An open-source information system for high-throughput plant phenotyping. *Plant Physiology*, *165*(2), 506–518. https://doi.org/10.1104/pp.113.233932

Kohli, S. K., Handa, N., Sharma, A., Kumar, V., Kaur, P., & Bhardwaj, R. (2017, December). Synergistic effect of 24-epibrassinolide and salicylic acid on photosynthetic efficiency and gene expression in *Brassica juncea* L. under Pb stress. *Turkish Journal of Biology*, *41*(6), 943–953. https://doi.org/10.3906/biy-1707-15

Komatsu, S., Kamal, A. H., & Hossain, Z. (2014, December 8). Wheat proteomics: Proteome modulation and abiotic stress acclimation. *Frontiers in Plant Science*, *5*, 684. https://doi.org/10.3389/fpls.2014.00684

Kosová, K., Vítámvás, P., Prášil, I. T., & Renaut, J. (2011, August 12). Plant proteome changes under abiotic stress – contribution of proteomics studies to understanding plant stress response. *Journal of Proteomics*, *74*(8), 1301–1322. https://doi.org/10.1016/j.jprot.2011.02.006

Kwak, M. J., Kong, H. G., Choi, K., Kwon, S. K., Song, J. Y., Lee, J., Lee, P. A., Choi, S. Y., Seo, M., & Lee, H. J. (2018, October 8). Rhizosphere microbiome structure alters to enable wilt resistance in tomato. *Nature Biotechnology*. https://doi.org/10.1038/nbt.4232

LaBaer, J., & Ramachandran, N. (2005). Protein microarrays as tools for functional proteomics. *Current Opinion in Chemical Biology*, *9*, 14–19. https://doi.org/10.1016/j.cbpa.2004.12.006

Lakhneko, O., Danchenko, M., Morgun, B., Kovac, A., Majerova, P., & Skultety, L. (2020). Comprehensive comparison of clinically relevant grain proteins in modern and traditional bread wheat cultivars. *International Journal of Molecular Sciences*, *21*(10), 3445. https://doi.org/10.3390/ijms21103445

Lappalainen, T., Sammeth, M., Friedländer, M. R., 't Hoen, P. A., Monlong, J., Rivas, M. A., Gonzàlez-Porta, M., Kurbatova, N., Griebel, T., & Ferreira, P. G. (2013, September 26). Transcriptome and genome sequencing uncovers functional variation in humans. *Nature*, *501*(7468), 506–511. https://doi.org/10.1038/nature12531

Laurindo, B. S., Laurindo, R. D. F., Fontes, P. P., Vital, C. E., Delazari, F. T., Baracat-Pereira, M. C., & da Silva, D. J. H. (2018). Comparative analysis of constitutive proteome between resistant and susceptible tomato genotypes regarding to late blight. *Functional & Integrative Genomics*, *18*, 11–21. https://doi.org/10.1007/s10142-017-0570-z

Le, D. T., Nishiyama, R., Watanabe, Y., Tanaka, M., Seki, M., Ham Le, H., Yamaguchi-Shinozaki, K., Shinozaki, K., & Tran, L. S. (2012). Differential gene expression in soybean leaf tissues at late developmental stages under drought stress revealed by genome-wide transcriptome analysis. *PLoS One*, *7*(11), e49522. https://doi.org/10.1371/journal.pone.0049522

Lee, H. J., Park, Y. J., Seo, P. J., Kim, J. H., Sim, H. J., Kim, S. G., & Park, C. M. (2015, December). Systemic immunity requires SnRK2.8-mediated nuclear import of NPR1 in Arabidopsis. *Plant Cell*, *27*(12), 3425–3438. https://doi.org/10.1105/tpc.15.00371

Lefevere, H., Bauters, L., & Gheysen, G. (2020, April 17). Salicylic acid biosynthesis in plants. *Frontiers in Plant Science*, *11*, 338. https://doi.org/10.3389/fpls.2020.00338

Li, T., Wang, Y. H., Liu, J. X., Feng, K., Xu, Z. S., & Xiong, A. S. (2019). Advances in genomic, transcriptomic, proteomic, and metabolomic approaches to study biotic stress in fruit crops. *Critical Reviews in Biotechnology*, *39*, 680–692. https://doi.org/10.1080/07388551.2019.1608153

Liebisch, F., Kirchgessner, N., Schneider, D., Walter, A., & Hund, A. (2015). Remote, aerial phenotyping of maize traits with a mobile multi-sensor approach. *Plant Methods*, *11*, 19. https://doi.org/10.1186/s13007-015-0048-8

Lindeberg, M. (2012). Genome-enabled perspectives on the composition, evolution, and expression of virulence determinants in bacterial plant pathogens. *Annual Review of Phytopathology*, *50*, 111–132. https://doi.org/10.1146/annurev-phyto-081211-173022

Liu, M. J., Zhao, J., Cai, Q. L., Liu, G. C., Wang, J. R., Zhao, Z. H., Liu, P., Dai, L., Yan, G., & Wang, W. J. (2014, October 28). The complex jujube genome provides insights into fruit tree biology. *Nature Communications*, *5*, 5315. https://doi.org/10.1038/ncomms6315

Lobos, G. A., Camargo, A. V., Del Pozo, A., Araus, J. L., Ortiz, R., & Doonan, J. H. (2017, December 22). Editorial: Plant phenotyping and phenomics for plant breeding. *Frontiers in Plant Science*, *8*, 2181. https://doi.org/10.3389/fpls.2017.02181

Lodha, T. D., Hembram, P., Tep, N., & Basak, J. (2013). Proteomics: A successful approach to understand the molecular mechanism of plant-pathogen interaction. *American Journal of Plant Sciences*, *4*, 1212–1226.

López-Gresa, M. P., Maltese, F., Bellés, J. M., Conejero, V., Kim, H. K., Choi, Y. H., & Verpoorte, R. (2010, January–February). Metabolic response of tomato leaves upon different plant-pathogen interactions. *Phytochemical Analysis*, *21*(1), 89–94. https://doi.org/10.1002/pca.1179

Luan, Y., Cui, J., Zhai, J., Li, J., Han, L., & Meng, J. (2015, June). High-throughput sequencing reveals differential expression of miRNAs in tomato inoculated with *Phytophthora infestans*. *Planta*, *241*(6), 1405–1416. https://doi.org/10.1007/s00425-015-2267-7

Luo, J., Xia, W., Cao, P., Xiao, Z., Zhang, Y., Liu, M., Zhan, C., & Wang, N. (2019, January 2). Integrated transcriptome analysis reveals plant hormones jasmonic acid and salicylic acid coordinate growth and defense responses upon fungal infection in poplar. *Biomolecules*, *9*(1), 12. https://doi.org/10.3390/biom9010012

Ma, H., Sheng, C., Qiao, L., Zhao, H., & Niu, D. (2020). A comparative proteomic approach to identify defence-related proteins between resistant and susceptible rice cultivars challenged with the fungal pathogen *Rhizoctonia solani*. *Plant Growth Regulation*, *90*, 73–88. https://doi.org/10.1007/s10725-019-00551-w

MacKenzie, T. D., Nie, X., & Singh, M. (2015). RT-PCR and real-time RT-PCR methods for the detection of potato virus Y in potato leaves and tubers. *Methods in Molecular Biology, 1236*, 13–26. https://doi.org/10.1007/978-1-4939-1743-3_2

Mani, S. D., Pandey, S., Govindan, M., Muthamilarasan, M., & Nagarathnam R. (2021, April). Transcriptome dynamics underlying elicitor-induced defense responses against Septoria leaf spot disease of tomato (*Solanum lycopersicum* L.). *Physiology and Molecular Biology of Plants, 27*(4), 873–888. https://doi.org/10.1007/s12298-021-00970-y

Massart, S., Perazzolli, M., Hofte, M., Pertot, I., & Ijakli, M. H. (2015, July 10). Impact of the omic technologies for understanding the modes of action of biological control agents against plant pathogens. *Biocontrol, 60*, 725–746. https://doi.org/10.1007/s10526=015-9686-z

McNeil, M. D., Bhuiyan, S. A., Berkman, P. J., Croft, B. J., & Aitken, K. S. (2018, May 24). Analysis of the resistance mechanisms in sugarcane during Sporisorium scitamineum infection using RNA-seq and microscopy. *PLoS One, 13*(5), e0197840. https://doi.org/10.1371/journal.pone.0197840

Mehta, A., Magalhaes, B. S., Djair, S. L., Erico, V. A. R., Luciano, S. P., Maria, G. F., Octavio, F. L., Paulo, H. A., & Thales, R. L. (2008). Rooteomics: The challenge of discovering plant defense-related proteins in roots. *Current Protein & Peptide Science, 9*(2), 108–116. https://doi.org/10.2174/138920308783955225

Mei, X., Lee, H. C., Diao, K. Y., Huang, M., Lin, B., Liu, C., Xie, Z., Ma, Y., Robson, P. M., Chung, M., Bernheim, A., Mani, V., Calcagno, C., Li, K., Li, S., Shan, H., Lv, J., Zhao, T., Xia, J., . . . Yang, Y. (2020). Artificial intelligence-enabled rapid diagnosis of patients with COVID-19. *Nature Medicine* [Epub ahead of print]. https://doi.org/10.1038/s41591-020-0931-3

Meng, H., Sun, M., Jiang, Z., Liu, Y., Sun, Y., Liu, D., Jiang, C., Ren, M., Yuan, G., Yu, W., Feng, Q., Yang, A., Cheng, L., & Wang, Y. (2021, January 12). Comparative transcriptome analysis reveals resistant and susceptible genes in tobacco cultivars in response to infection by *Phytophthora nicotianae*. *Scientific Reports, 11*(1), 809. https://doi.org/10.1038/s41598-020-80280-7

Mithöfer, A., & Boland, W. (2012). Plant defense against herbivores: Chemical aspects. *Annual Review of Plant Biology, 63*, 431–450. https://doi.org/10.1146/annurev-arplant-042110-103854

Moreira, X., Abdala-Roberts, L., Gols, R., & Francisco, M. (2018). Plant domestication decreases both constitutive and induced chemical defences by direct selection against defensive traits. *Scientific Reports, 8*, 12678. https://doi.org/10.1038/s41598-018-31041-0

Morkunas, I., & Ratajczak, L. (2014). The role of sugar signaling in plant defense responses against fungal pathogens. *Acta Physiologiae Plantarum, 36*, 1607–1619.

Morrell, P. L., Buckler, E. S., & Ross-Ibarra, J. (2011, December 29). Crop genomics: Advances and applications. *Nature Reviews Genetics, 13*(2), 85–96. https://doi.org/10.1038/nrg3097

Mou, Z., Fan, W., & Dong, X. (2003). Inducers of plant systemic acquired resistance regulate NPR1 function through redox changes. *Cell, 113*, 935–944.

Myhrvold, C., Freije, C. A., Gootenberg, J. S., Abudayyeh, O. O., Metsky, H. C., Durbin, A. F., Kellner, M. J., Tan, A. L., Paul, L. M., Parham, L. A., Garcia, K. F., Barnes, K. G., Chak, B., Mondini, A., Nogueira, M. L., Isern, S., Michael, S. F., Lorenzana, I., Yozwiak, N. L., . . . Sabeti, P. C. (2018, April 27). Field-deployable viral diagnostics using CRISPR-Cas13. *Science, 360*, 444–448. https://doi.org/10.1126/science.aas8836

Ng, D. W., Abeysinghe, J. K., & Kamali, M. (2018, November 24). Regulating the regulators: The control of transcription factors in plant defense signaling. *International Journal of Molecular Sciences, 19*(12), 3737. https://doi.org/10.3390/ijms19123737

Ninkovic, V., Rensing, M., Dahlin, I., & Markovic, D. (2019). Who is my neighbor? Volatile cues in plant interactions. *Plant Signaling & Behavior*, *14*(9), 1634993.

Notomi, T., Okayama, H., Masubuchi, H., Yonekawa, T., Watanabe, K., Amino, N., & Hase, T. (2000, June 15). Loop-mediated isothermal amplification of DNA. *Nucleic Acids Research*, *28*(12), e63. https://doi.org/10.1093/nar/28.12.e63

Núñez-Farfán, J., Fornoni, J., & Valverde, P. L. (2007). The evolution of resistance and tolerance to herbivores. *Annual Review of Ecology, Evolution, and Systematics*, *38*, 541–566. https://doi.org/10.1146/annurev.ecolsys.38.091206.095822

Nurnberger, T., & Kemmerling, B. (2006, November). Receptor protein kinases pattern recognition receptors in plant immunity. *Trends in Plant Science*, *11*(11), 519–522. https://doi.org/10.1016/j.tplants.2006.09.005

Oliver, J. D., Nilsson, L., & Kjelleberg, S. (1991, September). Formation of nonculturable Vibrio vulnificus cells and its relationship to the starvation state. *Applied and Environmental Microbiology*, *57*(9), 2640–2644. https://doi.org/10.1128/aem.57.9.2640-2644.1991

Oliver, S. G., Winson, M. K., Kell, D. B., & Baganz, F. (1998). Systematic functional analysis of the yeast genome. *Trends in Biotechnology*, *16*(9), 373–378. https://doi.org/10.1016/s0167-7799(98)01214-1

Pajerowska-Mukhtar, K. M., Emerine, D. K., & Mukhtar, M. S. (2013, July). Tell me more: Roles of NPRs in plant immunity. *Trends in Plant Science*, *18*(7), 402–411. https://doi.org/10.1016/j.tplants.2013.04.004

Parker, D., Beckmann, M., Zubair, H., Enot, D. P., Caracuel-Rios, Z., Overy, D. P., Snowdon, S., Talbot, N. J., & Draper, J. (2009, September). Metabolomic analysis reveals a common pattern of metabolic re-programming during invasion of three host plant species by Magnaporthe grisea. *Plant Journal*, *59*(5), 723–737.

Pentimone, I., Colagiero, M., Rosso, L. C., & Ciancio, A. (2020, May). Omics applications: Towards a sustainable protection of tomato. *Applied Microbiology and Biotechnology*, *104*(10), 4185–4195. https://doi.org/10.1007/s00253-020-10500-7

Pérez-Bueno, M. L., Pineda, M., Cabeza, F. M., & Barón, M. (2016, December 2). Multicolor fluorescence imaging as a candidate for disease detection in plant phenotyping. *Frontiers in Plant Science*, *7*, 1790. https://doi.org/10.3389/fpls.2016.01790

Peyraud, R., Dubiella, U., Barbacci, A., Genin, S., Raffaele, S., & Roby, D. (2017). Advances on plant-pathogen interactions from molecular toward systems biology perspectives. *Plant Journal*, *90*(4), 720–737.

Piasecka, A., Kachlicki, P., & Stobiecki, M. (2019, January 17). Analytical methods for detection of plant metabolomes changes in response to biotic and abiotic stresses. *International Journal of Molecular Sciences*, *20*(2).

Pieterse, C. M., & Dicke, M. (2007, December). Plant interactions with microbes and insects: From molecular mechanisms to ecology. *Trends in Plant Science*, *12*(12), 564–569. https://doi.org/10.1016/j.tplants.2007.09.004

Pieterse, C. M., Van der Does, D., Zamioudis, C., Leon-Reyes, A., & Van Wees, S. C. (2012). Hormonal modulation of plant immunity. *Annual Review of Cell and Developmental Biology*, *28*, 489–521. https://doi.org/10.1146/annurev-cellbio-092910-154055

Pombo, M. A., Zheng, Y., Fernandez-Pozo, N., Dunham, D. M., Fei, Z., & Martin, G.B. (2014). Transcriptomic analysis reveals tomato genes whose expression is induced specifically during effector-triggered immunity and identifies the Epk1 protein kinase which is required for the host response to three bacterial effector proteins. *Genome Biology*, *15*(10), 492. https://doi.org/10.1186/s13059-014-0492-1

Putney, S. D., Herlihy, W. C., & Schimmel, P. (1983, April 21). A new troponin T and cDNA clones for 13 different muscle proteins, found by shotgun sequencing. *Nature*, *302*(5910), 718–721. https://doi.org/10.1038/302718a0. PMID: 6687628.

Que, Y., Xu, L., Lin, J., Ruan, M., Zhang, M., & Chen, R. (2011). Differential protein expression in sugarcane during sugarcane-Sporisorium scitamineum interaction revealed by 2-DE and MALDI-TOF-TOF/MS. *Comparative and Functional Genomics*. https://doi.org/10.1155/2011/989016

Righetti, L., Bhandari, D. R., Rolli, E., Tortorella, S., Bruni, R., Dall'Asta, C., & Spengler, B. (2021). Unveiling the spatial distribution of aflatoxin B1 and plant defense metabolites in maize using AP-SMALDI mass spectrometry imaging. *The Plant Journal*, *106*(1), 185–199. https://doi.org/10.1111/tpj.15158

Robison, F. M., Turner, M. F., Jahn, C. E., Schwartz, H. F., Prenni, J. E., Brick, M. A., & Heuberger, A. L. (2018, September). Common bean varieties demonstrate differential physiological and metabolic responses to the pathogenic fungus Sclerotinia sclerotiorum. *Plant, Cell & Environment*, *41*(9), 2141–2154. https://doi.org/10.1111/pce.13176

Rodriguez, P. A., Rothballer, M., Chowdhury, S. P., Nussbaumer, T., Gutjahr, C., & Falter-Braun, P. (2019, June 3). Systems biology of plant-microbiome interactions. *Molecular Plant*, *12*(6), 804–821. https://doi.org/10.1016/j.molp.2019.05.006

Roessner, U., Luedemann, A., Brust, D., Fiehn, O., Linke, T., Willmitzer, L., & Fernie, A. (2001, January). Metabolic profiling allows comprehensive phenotyping of genetically or environmentally modified plant systems. *Plant Cell*, *13*(1), 11–29.

Rohrman, B. A., Leautaud, V., Molyneux, E., & Richards-Kortum, R. R. (2012). A lateral flow assay for quantitative detection of amplified HIV-1 RNA. *PLoS One*, *7*(9), e45611. https://doi.org/10.1371/journal.pone.0045611

Rojas, C. M., Senthil-Kumar, M., Tzin, V., & Mysore, K. S. (2014, February 10). Regulation of primary plant metabolism during plant-pathogen interactions and its contribution to plant defense. *Frontiers in Plant Science*, *5*, 17. https://doi.org/10.3389/fpls.2014.00017

Ruan, J., Zhou, Y., Zhou, M., Yan, J., Khurshid, M., Weng, W., Cheng, J., & Zhang, K. (2019, May 20). Jasmonic acid signaling pathway in plants. *International Journal of Molecular Sciences*, *20*(10), 2479. https://doi.org/10.3390/ijms20102479

Rubin, G. M., Yandell, J. R., Wortman, G. L., Gabor, C. R., Nelson, I. K., Hariharan, J. M., Cherry, F. W. (2000). Comparative genomics of the eukaryotes. *Science*, *287*(5461), 2204–2215.

Rubio, L., Herrero, J. R., Sarrio, J., Moreno, P., & Guerri, J. (2003). A new approach to evaluate relative resistance and tolerance of tomato cultivars to begomoviruses causing the tomato yellow leaf curl disease in Spain. *Plant Pathology*, *52*, 763–769. https://doi.org/10.1111/j.1365-3059.2003.00926

Sade, D., Shriki, O., Cuadros-Inostroza, A., Tohge, T., Semel, Y., Haviv, Y., Willmitzer, L., Fernie, A. R., Czosnek, H., & Brotman, Y. (2015). Comparative metabolomics and transcriptomics of plant response to Tomato yellow leaf curl virus infection in resistant and susceptible tomato cultivars. *Metabolomics*, *11*, 81–97.

Sakiyama, N. S., Ramos, H. C. C., Caixeta, E. T., & Pereira, M. G. (2014). Plant breeding with marker-assisted selection in Brazil. *Crop Breeding and Applied Biotechnology*, *14*, 54–60.

Sanchez-Flores, A., & Abreu-Goodger, C. (2014). A practical guide to sequencing genomes and transcriptomes. *Current Topics in Medicinal Chemistry*, *14*(3), 398–406.

Sanger, F., Nicklen, S., & Coulson, A. R. (1977). DNA sequencing with chain-terminating inhibitors. *Proceedings of the National Academy of Sciences*, *74*(12), 5463–5467. https://doi.org/10.1073/pnas.74.12.5463

Santini, A., & Ghelardini, L. (2015). Plant pathogen evolution and climate change. *CAB Reviews*, *10*, 1–35. https://doi.org/10.1079/PAVSNNR201510035

Sauter, H., Lauer, M., & Fritsch, H. (1991). Metabolic profiling of plants – a new diagnostic technique. In D. R. Baker, J. G. Fenyes, & W. K. Moberg (Hrsg.), *Synthesis and chemistry of agrochemicals II* (pp. 288–299). ACS Symposium Series 443. American Chemical Society.

Selvaraj, V., Maheshwari, Y., Hajeri, S., & Yokomi, R. (2019, September 5). A rapid detection tool for VT isolates of Citrus tristeza virus by immunocapture-reverse transcriptase loop-mediated isothermal amplification assay. *PLoS One*, *14*(9), e0222170. https://doi.org/10.1371/journal.pone.0222170

Seo, H. S., Song, J. T., Cheong, J. J., Lee, Y. H., Lee, Y. W., Hwang, I., Lee, J. S., & Choi, Y. D. (2001, April 10). Jasmonic acid carboxyl methyltransferase: A key enzyme for jasmonate-regulated plant responses. *Proceedings of the National Academy of Sciences of the United States of America*, *98*(8), 4788–4793. https://doi.org/10.1073/pnas.081557298

Seybold, H., Trempel, F., Ranf, S., Scheel, D., Romeis, T., & Lee, J. (2014, December). Ca2+ signalling in plant immune response: From pattern recognition receptors to Ca2+ decoding mechanisms. *New Phytologist*, *204*(4), 782–790. https://doi.org/10.1111/nph.13031

Sghaier-Hammami, B., Hammami, S., Baazaoui, N., Gómez-Díaz, C., & Jorrín-Novo, J. V. (2020). Dissecting the seed maturation and germination processes in the non-orthodox quercus ilex species based on protein signatures as revealed by 2-DE coupled to MALDI-TOF/TOF proteomics strategy. *International Journal of Molecular Sciences*, *21*(14), 4870. https://doi.org/10.3390/ijms21144870

Shao, H., Wang, H., & Tang, X. (2015, October 29). NAC transcription factors in plant multiple abiotic stress responses: Progress and prospects. *Frontiers in Plant Science*, *6*, 902. https://doi.org/10.3389/fpls.2015.00902

Sharma, S., & Dasgupta, I. (2012, April). Development of SYBR green I based real-time PCR assays for quantitative detection of Rice tungro bacilliform virus and Rice tungro spherical virus. *Journal of Virological Methods*, *181*(1), 86–92. https://doi.org/10.1016/j.jviromet.2012.01.018

Sharpton, T. J. (2014, June 16). An introduction to the analysis of shotgun metagenomic data. *Frontiers in Plant Science*, *5*, 209. https://doi.org/10.3389/fpls.2014.00209

Sheibani-Tezerji, R., Rattei, T., Sessitsch, A., Trognitz, F., & Mitter, B. (2015, September 8). Transcriptome profiling of the endophyte *Burkholderia phytofirmans* PsJN indicates sensing of the plant environment and drought stress. *mBio*, *6*(5), e00621. https://doi.org/10.1128/mBio.00621-15

Shuai, L., Liao, L., Duan, Z., Song, M., Wang, G., Wang, Z., & Zhang, M. (2020). Photosynthesis-related protein expression in sugarcane during sugarcane Fusarium verticillioides interaction revealed by iTRAQ. *Plant Growth Regulation*, *91*, 237–248.

Schlaeppi, K., Abou-Mansour, E., Buchala, A., & Mauch, F. (2010, June 1). Disease resistance of Arabidopsis to Phytophthora brassicae is established by the sequential action of indole glucosinolates and camalexin. *Plant Journal*, *62*(5), 840–851.

Schneider, D. J., & Collmer, A. (2010). Studying plant-pathogen interactions in the genomics era: Beyond molecular Koch's postulates to systems biology. *Annual Review of Phytopathology*, *48*, 457–479. https://doi.org/10.1146/annurev-phyto-073009-114411

Sicher, R. C., & Barnaby, J. Y. (2012). Impact of carbon dioxide enrichment on the responses of maize leaf transcripts and metabolites to water stress. *Physiologia Plantarum*, *144*(3), 238–253. https://doi.org/10.1111/j.1399-3054.2011.01555.x

Sim, G. K., Kafatos, F. C., Jones, C. W., Koehler, M. D., Efstratiadis, A., & Maniatis, T. (1979, December). Use of a cDNA library for studies on evolution and developmental expression of the chorion multigene families. *Cell*, *18*(4), 1303–1316. https://doi.org/10.1016/0092-8674(79)90241-1

Srivastava, S., Upadhyay, D. J., & Srivastava, A. (2020). Next-generation molecular diagnostics development by CRISPR/Cas tool: Rapid detection and surveillance of viral disease outbreaks. *Frontiers in Molecular Biosciences*, *7*, 582499. https://doi.org/10.3389/fmolb.2020.582499

Srivastava, V., Obudulu, O., Bygdell, J., Löfstedt, T., Rydén, P., Nilsson, R., Ahnlund, M., Johansson, A., Jonsson, P., Freyhult, E., Qvarnström, J., Karlsson, J., Melzer, M., Moritz, T., Trygg, J., Hvidsten, T. R., & Wingsle, G. (2013, December 17). OnPLS integration of transcriptomic, proteomic and metabolomic data shows multi-level oxidative stress responses in the cambium of transgenic hipI- superoxide dismutase Populus plants. *BMC Genomics*, *14*, 893. https://doi.org/10.1186/1471-2164-14-893

Sruthilaxmi, C. B., & Babu, S. (2020, July). Proteome responses to individual pathogens and abiotic conditions in rice seedlings. *Phytopathology*, *110*(7), 1326–1341. https://doi.org/10.1094/PHYTO-11-19-0425-R

Stagnari, F., Galieni, A., & Pisante, M. (2016). Drought stress effects on crop quality. In A. Parvaiz (Ed.), *Water stress and crop plants: A sustainable approach* (pp. 375–392). John Wiley & Sons, Ltd. https://doi.org/10.1002/9781119054450.ch23

Stotz, H. U., Harvey, P. J., Haddadi, P., Mashanova, A., Kukol, A., Larkan, N. J., Borhan, M. H., & Fitt, B. D. L. (2008, June 1). Genomic evidence for genes encoding leucine-rich repeat receptors linked to resistance against the eukaryotic extra- and intracellular Brassica napus pathogens *Leptosphaeria maculans* and *Plasmodiophora brassicae*. *PLoS One*, 13(6), e0198201. https://doi.org/10.1371/journal.pone.0198201. Erratum in: *PLoS One*, *13*(8), e0202143 (2018, August 7).

Sumner, L. W., Mendes, P., & Dixon, R. A. (2003, March). Plant metabolomics: Large-scale phytochemistry in the functional genomics era. *Phytochemistry*, *62*(6), 817–836. https://doi.org/10.1016/s0031-9422(02)00708-2

Sundin, G. W., Wang, N., Charkowski, A. O., Castiblanco, L. F., Jia, H., & Zhao, Y. (2016). Perspectives on the transition from bacterial phytopathogen genomics studies to applications enhancing disease management: From promise to practice. *Phytopathology*, *106*(10), 1071–1082. https://doi.org/10.1094/PHYTO-03-16-0117-FI

Sutcliffe, J. G., Milner, R. J., Bloom, F. E., & Lerner, R. A. (1982, August). Common 82-nucleotide sequence unique to brain RNA. *Proceedings of the National Academy of Sciences of the United States of America*, *79*(16), 4942–4946. https://doi.org/10.1073/pnas.79.16.4942

Swarbrick, P. J., Schulze-Lefert, P., & Scholes, J. D. (2006 June). Metabolic consequences of susceptibility and resistance (race-specific and broad-spectrum) in barley leaves challenged with powdery mildew. *Plant, Cell & Environment*, *29*(6), 1061–1076. https://doi.org/10.1111/j.1365-3040.2005.01472.x

Thomas, S., Karnik, S., Barai, R. S., Jayaraman, V. K., & Idicula-Thomas, S. (2010, January). CAMP: A useful resource for research on antimicrobial peptides. *Nucleic Acids Research*, *38*(Database issue), D774–D780. https://doi.org/10.1093/nar/gkp1021

Tian, B. Y., Cao, Y., & Zhang, K. Q. (2015, November 25). Metagenomic insights into communities, functions of endophytes, and their associates with infection by root-knot nematode, Meloidogyne incognita, in tomato roots. *Scientific Reports*, *5*, 17087. https://doi.org/10.1038/srep17087

Tyers, M., & Mann, M. (2003, March 13). From genomics to proteomics. *Nature*, *422*(6928), 193–197. https://doi.org/10.1038/nature01510

Van Emon, J. M. (2016). The omics revolution in agricultural research. *Journal of Agricultural and Food Chemistry*, *64*(1), 36–44. https://doi.org/10.1021/acs.jafc.5b04515

Varshney, R. K., Nayak, S. N., May, G. D., & Jackson, S. A. (2009, September). Next-generation sequencing technologies and their implications for crop genetics and breeding. *Trends in Biotechnology*, *27*(9), 522–530. https://doi.org/10.1016/j.tibtech.2009.05.006

Velculescu, V. E., Zhang, L., Vogelstein, B., & Kinzler, K. W. (1995, October 20). Serial analysis of gene expression. *Science*, *270*(5235), 484–487. https://doi.org/10.1126/science.270.5235.484

Virlet, N., Sabermanesh, K., Sadeghi-Tehran, P., & Hawkesford, M. J. (2016, February). Field Scanalyzer: An automated robotic field phenotyping platform for detailed crop monitoring. *Functional Plant Biology*, *44*(1), 143–153. https://doi.org/10.1071/FP16163

Wang, D., Weaver, N. D., Kesarwani, M., & Dong, X. (2005, May 13). Induction of protein secretory pathway is required for systemic acquired resistance. *Science*, *308*(5724), 1036–1040. https://doi.org/10.1126/science.1108791

Wang, S., Liu, L., Mi, X., Zhao, S., An, Y., Xia, X., Guo, R., & Wei, C. (2021, May). Multi-omics analysis to visualize the dynamic roles of defense genes in the response of tea plants to gray blight. *Plant Journal*, *106*(3), 862–875. https://doi.org/10.1111/tpj.15203

Wang, Y., Loake, G. J., & Chu, C. (2013, August 16). Cross-talk of nitric oxide and reactive oxygen species in plant programed cell death. *Frontiers in Plant Science*, *4*, 314. https://doi.org/10.3389/fpls.2013.00314

Wang, Z., Gerstein, M., & Snyder, M. (2009). RNA-Seq: A revolutionary tool for transcriptomics. *Nature Reviews Genetics*, *10*, 57–63.

Ward, E., Foster, S. J., Fraaije, B. A., & Mccartney, H. A. (2004). Plant pathogen diagnostics: Immunological and nucleic acid-based approaches. *Annals of Applied Biology*, *145*, 1–16. https://doi.org/10.3109/1040841X.2010.489892

Watanabe, S., Matsumoto, M., Hakomori, Y., Takagi, H., Shimada, H., & Sakamoto, A. (2014, April). The purine metabolite allantoin enhances abiotic stress tolerance through synergistic activation of abscisic acid metabolism. *Plant, Cell & Environment*, *37*(4), 1022–1036. https://doi.org/10.1111/pce.12218

Wenig, M., Ghirardo, A., Sales, J. H., Pabst, E. S., Breitenbach, H. H., Antritter, F., Weber, B., Lange, B., Lenk, M., Cameron, R. K., Schnitzler, J. P., & Vlot, A. C. (2019, August 23). Systemic acquired resistance networks amplify airborne defense cues. *Nature Communications*, *10*(1), 3813. https://doi.org/10.1038/s41467-019-11798-2

Winnenburg, R., Urban, M., Beacham, A., Baldwin, T. K., Holland, S., Lindeberg, M., Hansen, H., Rawlings, C., Hammond-Kosack, K. E., & Köhler, J. (2008, January). PHI-base update: Additions to the pathogen host interaction database. *Nucleic Acids Research*, *36*(Database issue), D572–D576. https://doi.org/10.1093/nar/gkm858

Woese, C. R., & Fox, G. E. (1977). Phylogenetic structure of the prokaryotic domain: The primary kingdoms. *Proceedings of the National Academy of Sciences of the United States of America*, *74*(11), 5088–5090. https://doi.org/10.1073/pnas.74.11.5088

Wu, Y., Zhang, D., Chu, J. Y., Boyle, P., Wang, Y., Brindle, I. D., De Luca, V., & Després, C. (2012, June 28). The Arabidopsis NPR1 protein is a receptor for the plant defense hormone salicylic acid. *Cell Reports*, *1*(6), 639–647. https://doi.org/10.1016/j.celrep.2012.05.008

Xie, H., Yang, D. H., Yao, H., Bai, G., Zhang, Y. H., & Xiao, B. G. (2016, January 15). iTRAQ-based quantitative proteomic analysis reveals proteomic changes in leaves of cultivated tobacco (*Nicotiana tabacum*) in response to drought stress. *Biochemical and Biophysical Research Communications*, *469*(3), 768–775. https://doi.org/10.1016/j.bbrc.2015.11.133

Yan, C., & Xie, D. (2015, December). Jasmonate in plant defence: Sentinel or double agent? *Plant Biotechnology Journal*, *13*(9), 1233–1240. https://doi.org/10.1111/pbi.12417

Yang, F., Zhang, Y., Huang, Q., Yin, G., Pennerman, K. K., Yu, J., Liu, Z., Li, D., & Guo, A. (2015). Analysis of key genes of jasmonic acid mediated signal pathway for defense against insect damages by comparative transcriptome sequencing. *Scientific Reports*, *5*, 16500. https://doi.org/10.1038/srep16500

Yang, J. K., Tong, Z. J., Fang, D. H., Chen, X. J., Zhang, K. Q., & Xiao, B. G. (2017). Transcriptomic profile of tobacco in response to *Phytophthora nicotianae* infection. *Scientific Reports*, *7*. https://doi.org/10.1038/s41598-017-00481-5

Yang, L., Lin, H., Takahashi, Y., Chen, F., Walker, M. A., & Civerolo, E. L. (2011). Proteomic analysis of grapevine stem in response to *Xylella fastidiosa* inoculation. *Physiological and Molecular Plant Pathology*, *75*, 90–99. https://doi.org/10.1016/j.pmpp.2010.11.002

Yao, D., Zhang, X., Zhao, X., Liu, C., Wang, C., Zhang, Z., Zhang, C., Wei, Q., Wang Q, Yan H., Li. F., & Su, Z. (2011, July). Transcriptome analysis reveals salt-stress-regulated biological processes and key pathways in roots of cotton (*Gossypium hirsutum* L.). *Genomics*, *98*(1), 47–55. https://doi.org/10.1016/j.ygeno.2011.04.007

Yu, Y., Qian, Y., Jiang, M., Xu, J., Yang, J., Zhang, T., Gou, L., & Pi, E. (2020, August 20). Regulation mechanisms of plant basic leucine zippers to various abiotic stresses. *Frontiers in Plant Science*, *11*, 1258. https://doi.org/10.3389/fpls.2020.01258

Zandalinas, S. I., Mittler, R., Balfagón, D., Arbona, V., Gómez-Cadenas, A. (2018, January). Plant adaptations to the combination of drought and high temperatures. *Physiologia Plantarum*, *162*(1), 2–12. https://doi.org/10.1111/ppl.12540

Zhang, H., Mao, R., Wang, Y., Zhang, L., Wang, C., Lv, S., Liu, X., Wang, Y., & Ji, W. (2019, November). Transcriptome-wide alternative splicing modulation during plant-pathogen interactions in wheat. *Plant Science*, *288*, 110160. https://doi.org/10.1016/j.plantsci.2019.05.023

Zhang, N., Zhou, S., Yang, D., & Fan Z. (2020, June 26). Revealing shared and distinct genes responding to JA and SA signaling in arabidopsis by meta-analysis. *Frontiers in Plant Science*, *11*, 908. https://doi.org/10.3389/fpls.2020.00908. Erratum in: *Frontiers in Plant Science*, *11*, 596349 (2020, September 17).

Zhang, W., Corwin, J. A., Copeland, D., Feusier, J., Eshbaugh, R., Chen, F., Atwell, S., & Kliebenstein, D. J. (2017b, November). Plastic transcriptomes stabilize immunity to pathogen diversity: The jasmonic acid and salicylic acid networks within the *Arabidopsis/Botrytis* pathosystem. *Plant Cell*, *29*(11), 2727–2752. https://doi.org/10.1105/tpc.17.00348

Zhang, W., Li, F., & Nie, L. (2010, February). Integrating multiple "omics" analysis for microbial biology: Application and methodologies. *Microbiology (Reading)*, *156*(Pt 2), 287–301. https://doi.org/10.1099/mic.0.034793-0

Zhang, X., Huang, C., Wu, D., Qiao, F., Li, W., Duan, L., Wang, K., Xiao, Y., Chen, G., Liu, Q., Xiong, L., Yang, W., & Yan, J. (2017a, March). High-throughput phenotyping and QTL mapping reveals the genetic architecture of maize plant growth. *Plant Physiology*, *173*(3), 1554–1564. https://doi.org/10.1104/pp.16.01516

Zhang, Y., Nan, J., & Yu, B. (2016, June 22). OMICS technologies and applications in sugar beet. *Frontiers in Plant Science*, *7*, 900. https://doi.org/10.3389/fpls.2016.00900

Zhang, Y., Song, L., Hou, L., Cao, Z., Vongsangnak, W., Zhu, G., Xu, Q., & Chen, G. (2021, August 5). Dual transcriptomic analyses unveil host-pathogen interactions between salmonella enterica serovar enteritidis and laying ducks (*Anas platyrhynchos*). *Frontiers in Microbiology*, *12*, 705712. https://doi.org/10.3389/fmicb.2021.705712

Zhang, Y., Zhang, Y. L., Chen, S. X., Yin, G. H., Yang, Z. Z., Lee, S., Liu, C. G., Zhao, D. D., Ma, Y. K., Song, F. Q., Bennett, J. W., & Yang, F. S. (2015, March 21). Proteomics of methyl jasmonate induced defense response in maize leaves against Asian corn borer. *BMC Genomics*, *16*(1), 224. https://doi.org/10.1186/s12864-015-1363-1

Zhao, C., Zhang, Y., Du, J., Guo, X., Wen, W., Gu, S., Wang, J., & Fan, J. (2019, June 3). Crop phenomics: Current status and perspectives. *Frontiers in Plant Science*, *10*, 714. https://doi.org/10.3389/fpls.2019.00714

6 Meta-Analysis of *Arabidopsis thaliana* Gene Expression Data Identifies Heavy Metal Stress Responsive Genes

Birendra Singh Yadav, Sandeep Kushwaha, Saurabh Yadav, and Ashutosh Mani

6.1 BACKGROUND

The source of cadmium is generally food, drinking water and smoking (Nawrot et al., 2006; Ursinyova & Hladikova, 2000; Verbruggen et al., 2009). A plant-based diet is also an important source of cadmium for humans (Peralta-Videa et al., 2009). The accumulation of cadmium sometimes also requires phytoremediation of toxic environments. Cadmium enters in cells through zinc transporter proteins (LeDuc & Terry, 2005).

Arsenic is a toxic heavy metal that is generally found in the environment and acts as potential carcinogen for humans (Jang et al., 2016; McGregor et al., 2000). Plants take in arsenic at low concentration in the form of As[V] and As[III]. As[III] (Quaghebeur & Rengel, 2003) is more toxic than As[V], as it reacts with sulfhydryl groups of enzymes and proteins that in turn inhibit cellular functions that may cause death (Ullrich-Eberius et al., 1989). As[V] is an analog of phosphate macronutrients, so it competes with Pi, which may lead to formation of an ADP-As complex, which is very unstable (Species & Hartley-Whitaker, 2015). Arsenate and arsenite both induce the activation of phytochelatin in plants (Schmöger et al., 2000). Cellular glutathione is depleted due to phytochelatin synthesis, which leads to decreased a quench capacity of reactive oxygen species (Hartley-Whitaker et al., 2001). Studies have shown that during exposure to As[V] stress, Cu/Zn SODs are down-regulated (Abercrombie et al., 2008).

Selenium is an essential trace element present in waste water and soils. It directly competes with sulphate (Minorsky, 2003; Ellis & Salt, 2003). Selenocysteine and

DOI: 10.1201/9781003153481-6

selenomethionine are formed due to the introduction of selenium to the sulphate assimilation pathway (Tamaoki et al., 2008). Enhanced exposure concentrations of selenium have been correlated with increased NO formation, affecting endogenous NO status within the root and stem in Arabidopsis plants (Lehotai, 2011).

Heavy metals like Cu, Zn and Ni are known to be essential micronutrients at low concentrations, but at higher concentrations they are known to be toxic (Munzuroglu & Geckil, 2002). Multiple pathways present in plants are mainly responsible for the uptake of heavy metal (Clemens, 2001). (Yadav et al., 2019) found that ten members of the bHLH transcription factor family regulate the different processes involved in abiotic stress by interacting with DNA. During exposure to heavy metals like Cu, Mn, Pb and Zn, Arabidopsis plants show a reduction in chlorophyll content and a gradual increase and subsequent decrease in carotenoid content, and also leads to accumulation of anthocyanin (Baek et al., 2012). Chickpea plants show enrichment in haloacid dehydrogenase, cellulose synthase, cinnamoyl CoA reducatase, beta-glucosidae 13, F-box protein, isoflavone hydroxylase and GDSL esterase lipase after exposure to Cr, Cd and As (Yadav et al., 2019). (Dixit & Dhankher, 2011) reported that the AtSAP10 gene is up-regulated in Arabidopsis plants during exposure to Ni, Mn, Zn and high temperature stress (Dixit & Dhankher, 2011). In rice, SAP (stress-associated protein) is induced by multiple abiotic stress factors like cold, drought, desiccation, wounding and heavy metals (Kanneganti & Gupta, 2008; Mukhopadhyay et al., 2004).

This study will include gene expression profiling of Arabidopsis plants based on treatment of different heavy metals (Cd, As. Se) at various concentrations. The microarray datasets used provide detailed information about the differential expressions observed during experimental conditions. The microarray datasets may be used to extract information that can provide a bigger picture of the whole system. Relevance of meta-analysis is increased with the high number of datasets used in the study. The analysis may be further useful to infer a common solution for different problems. The meta-analysis approach will be used to find out the genes that are generally up- or down-regulated during exposure to different heavy metals. This data analysis will be also helpful for finding interactive genes that are up-regulated and down-regulated with cellular localization, molecular function and gene network.

6.2 MATERIAL AND METHOD

6.2.1 Identification of Salt Stress Expression Dataset

Studies on abiotic stress on plants were identified by searching the PubMed database. The key words "*Arabidopsis thaliana*," "gene expression," "abiotic stress," and "microarray" were used to search. Microarray data are of two types: single color and double color. The first method measures the absolute expression of a gene during normal and experimental condition, while the second method measures the relative expression of a gene with respect to control. Abiotic stress (heavy metal) studies that use both types of microarray data – i.e., single color as well dual color – are available in databases. For performing meta-analysis, single color microarray data has been selected by searching the GEO database using "heavy metals," "gene expression" and Arabidopsis thaliana" as key words (Barrett et al., 2013). Microarray experiments with common platforms have been included in this study. GEO accession no., platform, no. of cases and control, sample type and gene expression data were extracted from each study.

6.2.2 Data Preprocessing

The gene expression data of the same platform are integrated; a probe with the same ID or a platform with a different ID should be uniform so that easy comparisons can be made with different microarray data. Normalization of integrated microarray data is done by calculating the Z score.

$$\text{Z score} = x_i - \bar{x} / \delta,$$

where x_i represents raw intensity data for each gene, $\bar{x}$ represents average gene intensity within a single experiment and δ represents standard deviation (SD) of all measured intensities.

The Z score allows the raw intensity value at a single scale.

6.2.3 Statistical Analysis

Differentially expressed genes (DEGs) were identified by comparing normal and treated samples. T-test was performed by calculating the average of control and treated sample. Log_2 of relative difference of treated and normal data was estimated for calculation of fold change. The genes with a higher fold change along with lower p value were shortlisted. Genes that have a fold change greater than two (> 2) were used for further studies (Tusher et al., 2001).

6.2.4 Functional Classification of Genes

Gene ontology (GO) was done to retrieve the biological significance of the DEGs of the *Arabidopsis thaliana* microarray data. The GENECODIS online server (Tusher et al., 2001) was used for this analysis. Along with GO, an enrichment analysis of DEGs based on KEGG was performed. GENECODIS was used to extract information about biological process (BP), molecular function (MF), cellular component (CC) and KEGG of DEGs obtained from statistical analysis of different microarray data (Tabas-Madrid et al., 2012).

6.2.5 Protein–Protein Interaction Network

The protein–protein interaction network was made by using GeneMANIA (Mostafavi et al., 2008). Cellular activities of DEGs with less than a 0.01 p-value, like growth, development, metabolism, differentiation and apoptosis, can be revealed at a molecular level by a PPI network. The preceding information plays an important role for defining or checking the cellular control mechanisms.

6.3 RESULTS

6.3.1 Short Overview of the Studies Included

The microarray data for a total of four expressions on *Arabidopsis thaliana* during exposure to different heavy metals include 28 samples, with 14 normal and 14 experimental conditions selected for the analysis (Table 6.1). All the expression data used in this study are results of experiments performed on the same platform.

6.3.2 Detected Gene Associated with Different Heavy Metals Exposure

Probe ID was unified for simultaneous comparison and analysis of all four expression datasets. The expression values were normalized by calculating the Z score at each spot of the expression matrix. A total of 7157 genes were found to be up-regulated, but not all are significant; 15652 genes are down-regulated. The top ten DEGs on the basis of p-value and fold change are shown in Table 6.2. For selecting significant up- and down-regulated genes after treatment with heavy metals, two tailed t-test and false discovery rate (FDR) (Benjamini & Hochberg, 1995) were calculated. The genes with fold change of more than two were selected and the p-value was calculated. The down-regulated gene with the lowest p-value 1.04×10^{-4} with a –2.644fold change was At1g25550, and the up-regulated gene with lowest p-value 0.000722 with a 2.191fold change was At3g62390.

6.3.3 Functional Annotation

GO category enrichment analysis was performed to gain molecular insights into DEGs in the system. Functional annotation and classification of genes was done by gene ontology. Gene ontology categorizes genes into three groups: biological process, cellular component and molecular function. GENECODIS was used for

TABLE 6.1
Properties of Previous Microarray Studies

GEO ID	Sample Count (case:control)	Platform	Sample Source	Tissue
GSE35869	1:1	GPL198/ATH1–121501/Affymetrix	*In vivo*	Seedlings
GSE22114	3:3	GPL198/ATH1–121501/Affymetrix	*In vivo*	Roots
GSE31977	6:6	GPL198/ATH1–121501/Affymetrix	*In vivo*	Roots
GSE9311	4:4	GPL198/ATH1–121501/Affymetrix	*In vivo*	Roots/Shoots

TABLE 6.2
The Top Ten Most Significantly Up- or Down-Regulated DEGs

Gene ID	Probe ID	P-Value	Fold Change	Function
Down-Regulated Genes				
255734_at	At1g25550	1.04E-05	−2.64452	myb-like transcription factor family protein
262399_at	At1g49500	9.18E-05	−3.34521	Unknown protein
261711_at	At1g32700	9.59E-05	−6.38744	Transcription factor family protein
252040_at	At3g52060	0.00011	−4.29402	Core-2/I-branching beta-1,6-N acetylglucosaminyltransferase family protein
251603_at	At3g57760	0.000126	−2.12841	Protein kinase superfamily protein
256255_at	At3g11280	0.000135	−6.20203	MYB family transcription factor
250105_at	At5g16630	0.00016	−2.03578	DNA repair protein Rad4 family
257615_at	At3g26510	0.000182	−2.53355	Octicosapeptide/Phox/Bem1p family protein
257517_at	At3g16330	0.000192	−3.75187	Unknown protein
253749_at	At4g29080	0.000214	−2.38431	Phytochrome-associated protein 2
Up-Regulated Gene				
251219_at	At3g62390	0.000722	2.191004882	TRICHOME BIREFRINGENCE-LIKE 6
260812_at	At1g43650	0.000754	2.150356657	EamA-like transporter family protein
264183_at	At1g65380	0.000805	3.211213481	Leucine-rich repeat (LRR) family protein
261576_at	At1g01070	0.000951	2.15827155	EamA-like transporter family protein
263386_at	At2g40150	0.00096	2.368248201	Wall associated kinase1
248963_at	At5g45700	0.000997	2.382473455	Haloacid dehalogenase-like hydrolase (HAD) superfamily protein
248656_at	At5g48460	0.001067	2.612573905	Actin binding calponin homology (CH) domain-containing protein
255460_at	At4g02800	0.001089	2.608391448	Unknown protein
255962_at	At1g22335	0.001095	2.970521123	Unknown protein
257715_at	At3g12750	0.001108	2.606101423	Zinc transporter 1 precursor

categorization of genes into biological and cellular components, for which a significance level of $p < 0.01$ was tested against the whole gene with GO annotation (Figures 6.1 and 6.2).

6.3.4 Protein–Protein Interaction Network Construction

The total number of down-regulated genes with a p-value of less than 0.05 is 2215, and the number of up-regulated genes with the same p-value threshold is 1663. In PPI, genes are represented by nodes and interactions are shown by edges. The

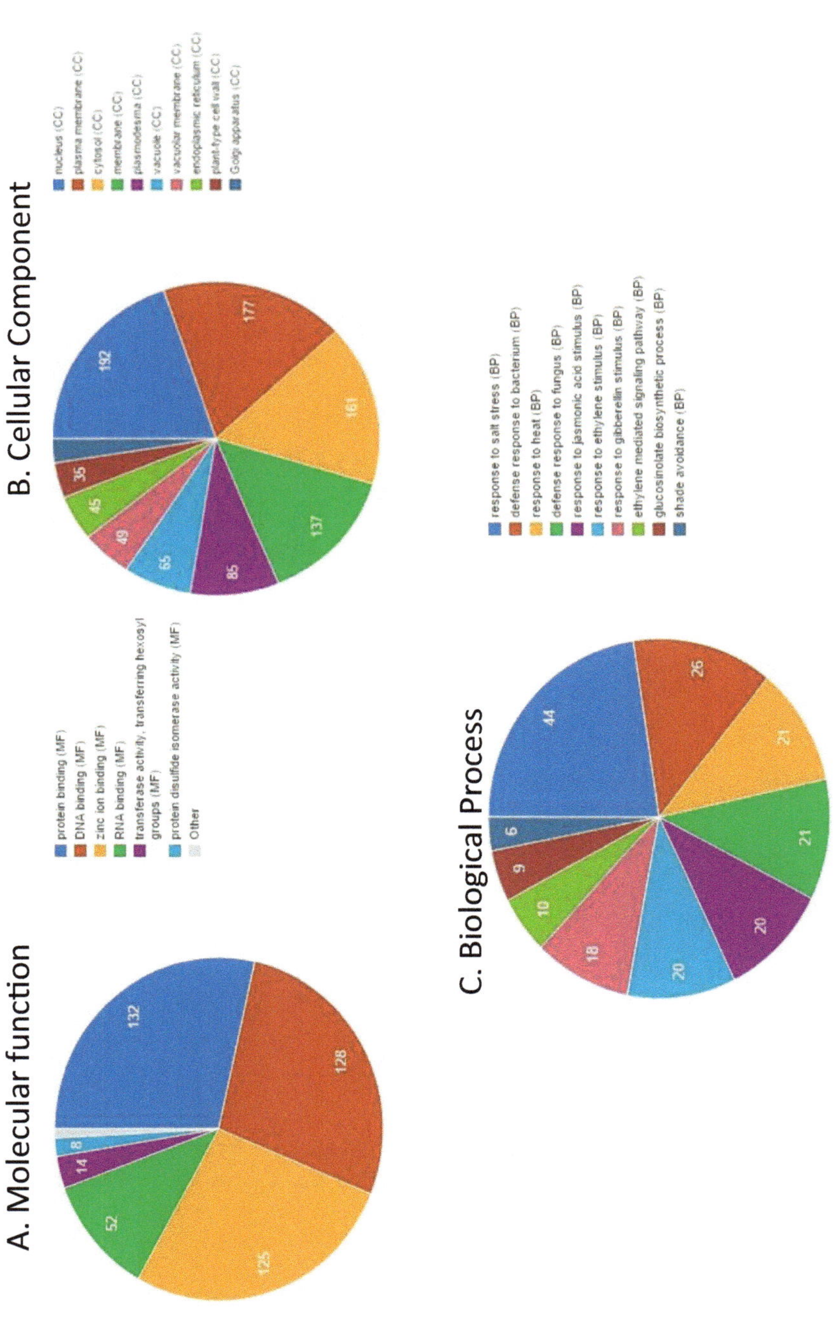

FIGURE 6.1 Gene ontology of differentially expressed down-regulated genes (p-value < 0.01).

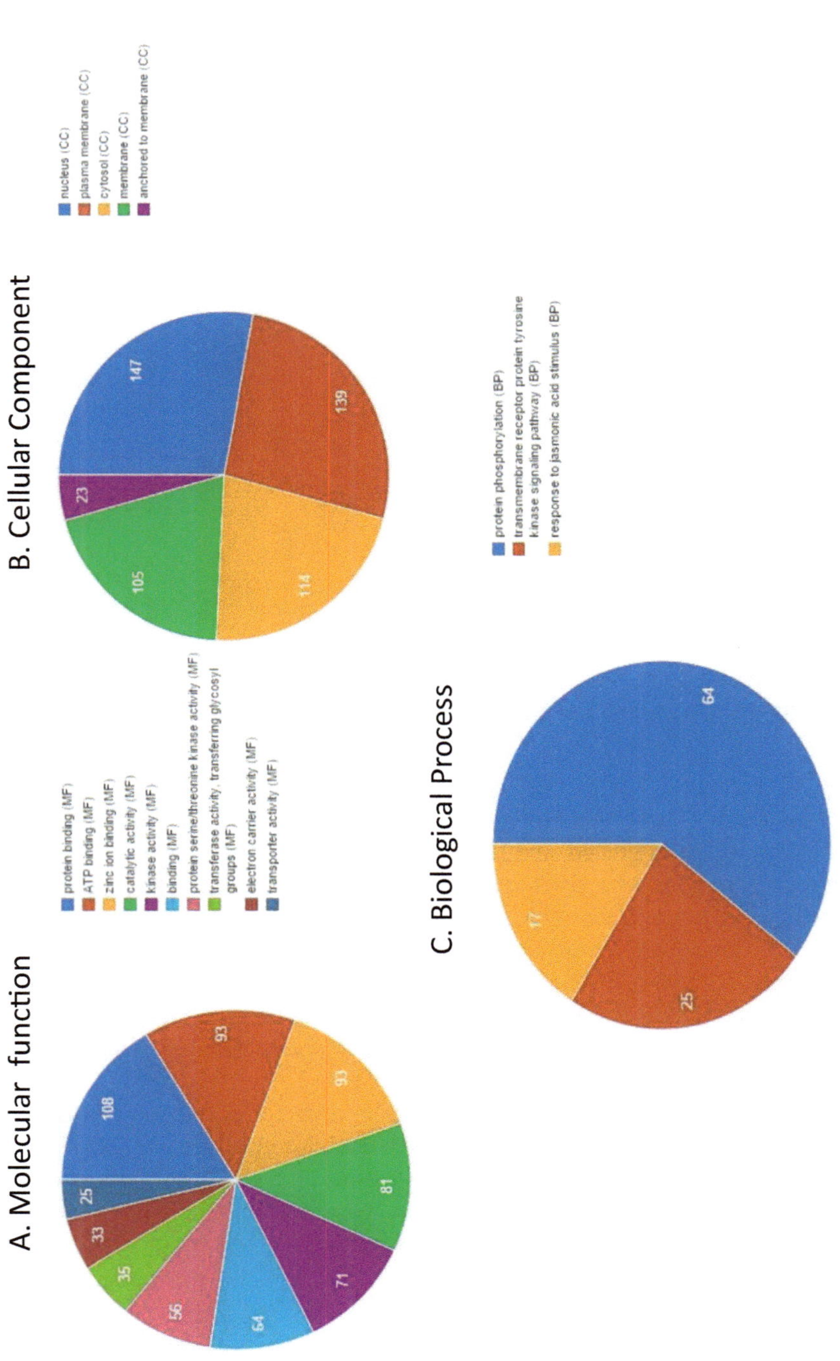

FIGURE 6.2 Gene ontology of differentially expressed up-regulated genes (p-value < 0.01).

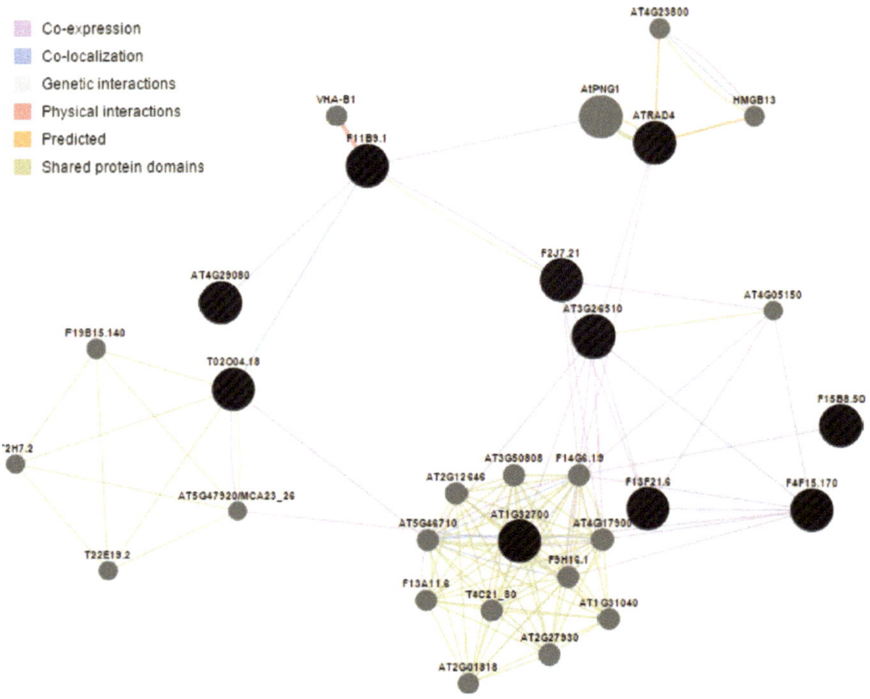

FIGURE 6.3 Gene network of top ten down-regulated genes.

protein/genes that are co-expressed and co-localized and have physical interactions with up-regulated and down-regulated gene are shown. The top ten up-regulated gene and down-regulated genes were selected for construction of PPI, and information regarding expression, localization and function was inferred. GeneMANIA was used to construct the PPI of DEGs, which is shown in Figures 6.3 and 6.4. At1g49500 has similarity to plant protein and is located in the endomembrane system and At3g16330 is an unknown protein that also has similarities to plant protein, while the eight other up-regulated proteins in the top ten are well known, as is their function.

At4g02800 has similarities with archaea, bacteria, fungi, plant and viruses and is located in the chloroplast, while the function of this gene and At1g22335 is still unknown. A total of 127 and 181 links were generated, with 20 different topological proteins related to the top ten up-regulated and down-regulated query genes respectively. These links represent various interactions with topological proteins in different ways (Table 6.3).

6.4 DISCUSSION AND CONCLUSION

Heavy metals become toxic at higher concentrations in plants. A comparative analysis of different heavy metal (As, Cd, Se) exposure on Arabidopsis plants has

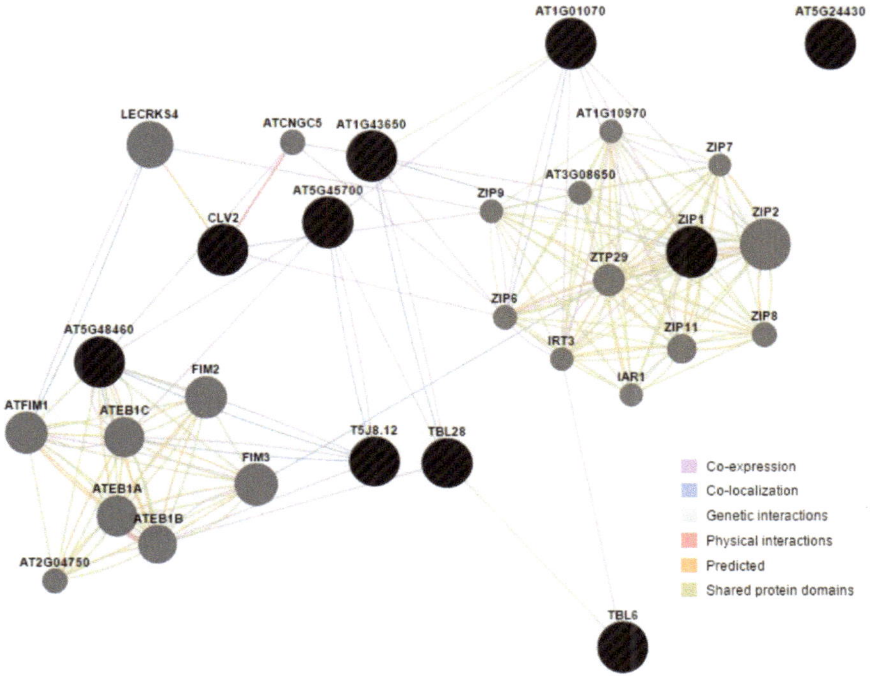

FIGURE 6.4 Gene network of top ten up-regulated genes.

TABLE 6.3
Showing the Way of Interaction and Percentage of Interacting Protein

Way of Interaction	Interacting Protein (%)
Physical interactions	39.02%
Predicted	30.87%
Shared protein domains	14.60 %
Co-expression	11.77 %
Co-localization	0.41

been performed to reveal the mechanism of heavy metal tolerance. In this chapter, we combined the microarray dataset of a different heavy metal treated matrix on Arabidopsis plants for meta-analysis that combines differentially expressed genes (DEGs) and statistically significant genes to further explore GO enrichment analysis along with protein–protein interaction (PPI).

Four publicly available GEO datasets were used to identify common biological mechanisms involved in heavy metal tolerance. A total of 1036 genes across the studies were consistently expressed differentially in Arabidopsis plants during heavy metal exposure (579 up-regulated and 457 down-regulated). The up-regulated gene with the lowest p-value ($p=7.2\times10^{-4}$) was TBL6 (Trichome Birefringence like-6) of unknown protein function, while the down-regulated gene with the lowest p-value ($p=1.039\times10^{-5}$) was a myb-like transcription family protein, which on over-expression increased the hypocotyl elongation in Arabidopsis plants and MYBH expression, triggered by challenging environmental conditions that lead to an increase in PIF (phytochrome interacting factor) accumulation (Kwon et al., 2013). One gene that was down-regulated was 262399_at, of unknown function protein. Other genes that were down-regulated included 261711_at, 252040_at, 251603_at, 256255_at, 250105_at, 257615_at, 257517_at and 253749_at, with functions of the transcription factor family protein, Core-2/I-branching beta-1,6-N acetylglucosaminyltransferase family protein, protein kinase superfamily protein, MYB family transcription factor, DNA repair protein Rad4 family, octicosapeptide/Phox/Bem1p family protein, unknown protein and phytochrome-associated protein 2 respectively.

The up-regulated genes are 260812_at, 264183_at, 261576_at, 263386_at, 248963_at, 248656_at, 255460_at, 255962_at and 257715_at, with their functions EamA-like transporter family protein, leucine-rich repeat (LRR) family protein, EamA-like transporter family protein, wall-associated kinase1, haloacid dehalogenase-like hydrolase (HAD) superfamily protein, actin-binding calponin homology (CH) domain-containing protein, unknown protein, unknown protein and zinc transporter 1 precursor.

GO category enrichment analysis was performed for prediction of the biological roles of DEGs. The down-regulated GO terms for molecular function were protein binding, DNA binding and zinc binding proteins, while the related biological functions were process response to salt stress, defense response to bacterium and responses to heat; for the cellular component, the enriched GO enrichment terms were nucleus, plasma membrane and cytosolic region. In the case of up-regulated DEGs, GO enrichment terms for molecular function were protein binding, ATP binding genes and zinc ion bindings; in biological process they were protein phosphorylation, trans membrane receptor protein and tyrosine kinase protein, while biological function and cellular component are same as up-regulated genes.

In order to investigate the interaction among several co-expressed and co-localized proteins, a protein–protein interaction network was constructed. The down-regulated gene myb-like transcription factor protein was found to exhibit physical interaction with calcium-dependent protein kinase and was co-expressing with Homedomain like superfamily protein, myb transcription factor protein, GLK-2 and ATCDPK-2 (calcium-dependent protein kinase–2), while up-regulated genes with lower p-values also had more than one interacting protein like co-localized and co-expressed.

Despite limitations like variation in data due to collection of mRNA from different sources, and more than one type of heavy metals being used in the different studies drawn on by this chapter, this may be first type of study to enable us to find the common mechanism involved during different heavy metal exposure to plants.

REFERENCES

Abercrombie, J. M., Halfhill, M. D., Ranjan, P., Rao, M. R., Saxton, A. M., Yuan, J. S., & Stewart, C. N. (2008). Transcriptional responses of Arabidopsis thaliana plants to As (V) stress. *BMC Plant Biology*, *8*, 87. https://doi.org/10.1186/1471-2229-8-87

Baek, S.-A., Han, T., Ahn, S.-K., Kang, H., Cho, M. R., Lee, S.-C., & Im, K.-H. (2012). Effects of heavy metals on plant growths and pigment contents in Arabidopsis thaliana. *The Plant Pathology Journal*, *28*(4), 446–452.

Barrett, T., Wilhite, S. E., Ledoux, P., Evangelista, C., Kim, I. F., Tomashevsky, M., Marshall, K. A., Phillippy, K. H., Sherman, P. M., Holko, M., Yefanov, A., Lee, H., Zhang, N., Robertson, C. L., Serova, N., Davis, S., & Soboleva, A. (2013). NCBI GEO: Archive for functional genomics data sets – update. *Nucleic Acids Research*, *41*(Database issue), D991–D995. https://doi.org/10.1093/nar/gks1193

Benjamini, Y., & Hochberg, Y. (1995). Controlling the false discovery rate: A practical and powerful approach to multiple testing. *Journal of the Royal Statistical Society: Series B (Methodological)*, *57*(1), 289–300.

Clemens, S. (2001). Molecular mechanisms of plant metal tolerance and homeostasis. *Planta*, *212*(4), 475–486. https://doi.org/10.1007/s004250000458

Dixit, A. R., & Dhankher, O. P. (2011). A novel stress-associated protein "AtSAP10" from Arabidopsis thaliana confers tolerance to nickel, manganese, zinc, and high temperature stress. *PLoS One*, *6*(6), e20921. https://doi.org/10.1371/journal.pone.0020921

Ellis, D. R., & Salt, D. E. (2003). Plants, selenium and human health. *Current Opinion in Plant Biology*, *6*(3), 273–279. https://doi.org/10.1016/s1369-5266(03)00030-x

Hartley-Whitaker, J., Ainsworth, G., Vooijs, R., Ten Bookum, W., Schat, H., & Meharg, A. A. (2001). Phytochelatins are involved in differential arsenate tolerance in Holcus lanatus. *Plant Physiology*, *126*(1), 299–306. https://doi.org/10.1104/pp.126.1.299

Jang, Y., Somanna, Y., & Kim, H. (2016). Source, distribution, toxicity and remediation of arsenic in the environment – a review. *International Journal of Applied Environmental Sciences*, *11*(2), 559–581.

Kanneganti, V., & Gupta, A. K. (2008). Overexpression of OsiSAP8, a member of stress associated protein (SAP) gene family of rice confers tolerance to salt, drought and cold stress in transgenic tobacco and rice. *Plant Molecular Biology*, *66*(5), 445–462. https://doi.org/10.1007/s11103-007-9284-2

Kwon, Y., Kim, J. H., Nguyen, H. N., Jikumaru, Y., Kamiya, Y., Hong, S.-W., & Lee, H. (2013). A novel Arabidopsis MYB-like transcription factor, MYBH, regulates hypocotyl elongation by enhancing auxin accumulation. *Journal of Experimental Botany*, *64*(12), 3911–3922.

LeDuc, D. L., & Terry, N. (2005). Phytoremediation of toxic trace elements in soil and water. *Journal of Industrial Microbiology and Biotechnology*, *32*(11–12), 514–520.

Lehotai, N. (2011). The effect of selenium (Se) on development and nitric oxide levels in Arabidopsis thaliana seedlings. *Acta Biologica Szegediensis*, *55*(1), 105–107.

McGregor, D. B., Baan, R. A., Partensky, C., Rice, J. M., & Wilbourn, J. D. (2000). Evaluation of the carcinogenic risks to humans associated with surgical implants and other foreign bodies – a report of an IARC monographs programme meeting: International agency for research on cancer. *European Journal of Cancer*, *36*(3), 307–313. https://doi.org/10.1016/s0959-8049(99)00312-3

Minorsky, P. V. (2003). The hot and the classic. *Plant Physiology*, *132*(2), 726–727.

Mostafavi, S., Ray, D., Warde-Farley, D., Grouios, C., & Morris, Q. (2008). GeneMANIA: A real-time multiple association network integration algorithm for predicting gene function. *Genome Biology*, *9*(Suppl 1), S4. https://doi.org/10.1186/gb-2008-9-s1-s4

Mukhopadhyay, A., Vij, S., & Tyagi, A. K. (2004). Overexpression of a zinc-finger protein gene from rice confers tolerance to cold, dehydration, and salt stress in transgenic tobacco. *Proceedings of the National Academy of Sciences of the United States of America*, *101*(16), 6309–6314. https://doi.org/10.1073/pnas.0401572101

Munzuroglu, O., & Geckil, H. (2002). Effects of metals on seed germination, root elongation, and coleoptile and hypocotyl growth in Triticum aestivum and Cucumis sativus. *Archives of Environmental Contamination and Toxicology*, *43*(2), 203–213.

Nawrot, T., Plusquin, M., Hogervorst, J., Roels, H. A., Celis, H., Thijs, L., Vangronsveld, J., Van Hecke, E., & Staessen, J. A. (2006). Environmental exposure to cadmium and risk of cancer: A prospective population-based study. *Lancet Oncology*, *7*(2), 119–126. https://doi.org/10.1016/s1470-2045(06)70545-9

Peralta-Videa, J. R., Lopez, M. L., Narayan, M., Saupe, G., & Gardea-Torresdey, J. (2009). The biochemistry of environmental heavy metal uptake by plants: Implications for the food chain. *The International Journal of Biochemistry & Cell Biology*, *41*(8–9), 1665–1677. https://doi.org/10.1016/j.biocel.2009.03.005

Quaghebeur, M., & Rengel, Z. (2003). The distribution of arsenate and arsenite in shoots and roots of Holcus lanatus is influenced by arsenic tolerance and arsenate and phosphate supply. *Plant Physiology*, *132*(3), 1600–1609. https://doi.org/10.1104/pp.103.021741

Schmöger, M. E., Oven, M., & Grill, E. (2000). Detoxification of arsenic by phytochelatins in plants. *Plant Physiology*, *122*(3), 793–801. https://doi.org/10.1104/pp.122.3.793

Species, P., & Hartley-Whitaker, J. (2015). Review Tansley Arsenic and metabolism in uptake and nonresistant resistant arsenic plant species. *New Phytologist*, *154*, 29–43.

Tabas-Madrid, D., Nogales-Cadenas, R., & Pascual-Montano, A. (2012). GeneCodis3: A non-redundant and modular enrichment analysis tool for functional genomics. *Nucleic Acids Res*, *40*(Web Server issue), W478–W483. https://doi.org/10.1093/nar/gks402

Tamaoki, M., Freeman, J. L., & Pilon-Smits, E. A. (2008). Cooperative ethylene and jasmonic acid signaling regulates selenite resistance in Arabidopsis. *Plant Physiology*, *146*(3), 1219–1230. https://doi.org/10.1104/pp.107.110742

Tusher, V. G., Tibshirani, R., & Chu, G. (2001). Significance analysis of microarrays applied to the ionizing radiation response. *Proceedings of the National Academy of Sciences of the United States of America*, *98*(9), 5116–5121. https://doi.org/10.1073/pnas.091062498

Ullrich-Eberius, C., Sanz, A., & Novacky, A. (1989). Evaluation of arsenate-and vanadate-associated changes of electrical membrane potential and phosphate transport in Lemna gibba G1. *Journal of Experimental Botany*, 119–128.

Ursinyova, M., & Hladikova, V. (2000). Cadmium in the environment of Central Europe. In *Trace metals in the environment* (Vol. 4, pp. 87–107). Elsevier.

Verbruggen, N., Hermans, C., & Schat, H. (2009). Mechanisms to cope with arsenic or cadmium excess in plants. *Current Opinion in Plant Biology*, *12*(3), 364–372. https://doi.org/10.1016/j.pbi.2009.05.001

Yadav, B. S., & Mani, A. (2019). Analysis of bHLH coding genes of *Cicer arietinum* during heavy metal stress using biological network. *Physiology and Molecular Biology of Plants: An International Journal of Functional Plant Biology*, *25*(1), 113–121. https://doi.org/10.1007/s12298-018-0625-1

Yadav, B. S., Singh, S., Srivastava, S., & Mani, A. (2019). Analysis of chickpea gene co-expression networks and pathways during heavy metal stress. *Journal of Biosciences*, *44*(4), 99.

7 Impact of Genomics Tools and Techniques on Crop Improvement

Shrishti Naidu, Saurav Suman, Amit Roy, Amrita Chakraborty, Indrakant K. Singh, and Archana Singh

7.1 INTRODUCTION

Advancements in genomics-related tools and techniques, from discovering vectors to genome sequencing methods and next-generation sequencing (NGS) and other biotechnological interventions, have opened up the possibilities for identifying genes/loci related to economically important traits in both model and non-model crop plants. Genomics studies involve deciphering the entire genetic complement of a species by employing recombinant DNA technology, DNA sequencing techniques, and *in silico* tools to sequence, assemble, and analyze the complete genomes of various organisms. Whole-genome sequencing provides a means to identify the complete set of genes and recognize the regulatory genetic factors, which control the crucial traits that can be explored further for their utilization in crop improvement. Since whole genome sequences are available for many crop plants, the focus of current plant scientists has moved from "how to sequence a genome" to "how to use the genomic information for crop improvement". Analysis of whole-genome sequences started with bioinformatic tools and techniques. The wide array of bioinformatic tools offer acquisition and organization of data at data repositories, which will help to understand the model plant systems and identify potential candidate genes, single nucleotide polymorphism (SNPs), simple sequence repeats (SSRs), and discovery of novel molecular/genetic markers. This information can be further utilized for molecular breeding and accelerated crop breeding. With the recent development of genomics techniques, it is feasible to identify all or most of the genes in a plant, and one can start to comprehend the genetic features and connections that promote crop productivity. These also contribute to the production of commercially significant crops and bring about a new age of research and innovation. The advent of novel sequencing technology and bioinformatics has improved the decoding of massive and diverse genomes, opening new avenues for crop improvements. Bioinformatics also has tremendous applications in transcriptomics, functional genomics, gene identification, protein–protein interaction, system biology, etc. (Bunnik & Le Roch, 2013).

DOI: 10.1201/9781003153481-7

Recent progress in sequencing methods led to the discovery of NGS, which obtains the sequence data of the genomes, transcriptomes, and expression profiles of the complete set of genes differentially expressed under particular physiological conditions. NGS assists in identifying the determining factors of a specific trait that can be further utilized for plant breeding or genetic modification (Edwards & Batley, 2010). NGS also helps to identify potential candidate genes, SNPs, mutations, and epigenetic changes that regulate traits transmitted from one generation to another (Bevan & Uauy, 2013). Subsequently, Genome wide association studies (GWAS) came into existence, which is a powerful tool for studying quantitative and complex characters related to economically important traits. Nowadays, it has become one of the most popular tools for investigating genotype–phenotype associations in many crop plants. Additionally, functional genomics tools such as overexpression, gene silencing, and one of the recently developed techniques, genome editing by CRISPR/Cas9, have been utilized in a couple of crop plants (Ricroch et al., 2017) for a variety of features such as yield improvement, biofortification, and environmental stress management as well as for functional characterization of genes.

This chapter highlights different approaches to whole-genome sequencing of economically important crops, such as whole-genome sequencing (WGS) and other relevant tools of genomics like GWAS, NGS, and biotechnological interventions, with an emphasis on overexpression, gene silencing, and CRISPR/Cas9 genome editing techniques and their applications. Bioinformatics approaches in various fields of biology and their application to crop improvement have also been considered. Despite having a broad application of genome sequencing, it remains challenging to utilize the voluminous data/information for knowledge that can be further exploited for crop improvement programs.

7.2 WHOLE GENOME SEQUENCING OF CROP PLANTS AND ITS ROLE IN IDENTIFICATION OF GENETIC FACTORS REGULATING AGRONOMIC TRAITS

To improve crop yield and understand plant growth and associated properties, it is critical to acquire entire genetic information/whole-genome sequences (WGS). Until 2010, bacterial artificial chromosome (BAC) and generating physical maps remained the most common WGS method for prominent crop genomes, including rice, poplar, and corn. The rice genome sequencing project has provided complete sequences of solo BACs that were processed into physical maps and subsequently linked to genetic maps (Schnable et al., 2009a; Tuskan et al., 2006b); however, sequences of solo BACs in maize were not completed (Schnable et al., 2009a; Tuskan et al., 2006a). With powerful techniques like *ARACHNE*, whole genome shotgun (WGS) reads of libraries of arbitrary segments of variable dimensions and *BAC terminal sequence* (BTS) are constructed for plant lineages, including soybean, grapevine, millet, and poplar (Batzoglou et al., 2002; Hood et al., 2004). Sorghum is the first crop plant to have its whole genome decoded using

WGS sequence assembly (Paterson et al., 2009; Schnable et al., 2009a; Tuskan et al., 2006b) and has been checked for righteousness post-genome sequencing utilizing high-density genomic mapping. The contiguousness of assemblies has differed depending on genomic makeup and size. Alignment to BAC sequences has been used to obtain high contiguity in alfalfa and potato (Velasco et al., 2010; Young et al., 2011). The sequencing of polyploid genomes such as *Brassica* has been described as the most challenging genome to sequence for massive assembly since three whole-genome duplications proceeded by partial diploidizations have been witnessed (Town et al., 2006). In contrast to its near cousin *Arabidopsis*, polyploidy has resulted in substantial structural diversity and gene copy number variability in *Brassica* species. The *Brassica rapa* genomic sequencing stayed unusual. However, *Brassica* chromosomal fragment alignment to the genome of *Arabidopsis* was shown to be extremely valuable for modern genetic investigations and evaluation (Bancroft et al., 2011).

Different combinations of sequencing technologies, all of which utilize WGS methods, have been effectively implemented for genomic sequencing of not only trees (such as date palm, cacao, apple) but also other fruits (like strawberry) and vegetables (such as potato and Chinese cabbage), as well as fodder crops such as alfalfa and its relatives in 2011 (Velasco et al., 2010; Young et al., 2011). The *Medicago* and *Solanum lycopersicum* crop initiatives began in the BAC-dependent Sanger sequencing period and were eventually finished using next-generation sequencing technologies. Until 2012, the genomic sequences of around 12 crop plants were released. Advanced sequencing platforms (like Illumina and oxford nanopore sequencing), assembling approaches, and user-friendly methodologies contributed to enormous data generation in the recent past and will continue to do so in future.

Just paired-end reads were used to decode the date palm genome, which remained fragmentary. Even though this was due to repeat or duplication structure, it is indistinct that using *mate-pair libraries* of varying lengths gives precisely spaced sets of sequence reads, which braces contiguity between all intermediate genomes of around 1,000 Mb, such as pigeon pea, diploid cotton, citrus, chickpea, and banana genomes (D'Hont et al., 2012; Varshney et al., 2011a, 2013; M. Wang et al., 2012; Xu et al., 2013). The gene-based assembling from wheat and barley was seen to be appropriate for producing genomic maps for map-based cloning and marker-assisted breeding, as well as for assembling genetic markers. BACs in the physical map were multiplex-sequenced utilizing Illumina techniques to boost gene covering and contiguity of the barley genome. Consequently, chromosomal assemblages with lesser gaps and better-ordered genes were generated. This positioned barley as the Triticeae family's predominant genetic template and genomic model (Allen et al., 2011). Few of the coniferous trees *Picea abies* (Norway spruce) and *Pinus taeda* (loblolly pine) are industrially useful, but they have relatively giant genomes (about 20,000–24,000 Mb, etc.) (Nystedt et al., 2013). WGS methodologies comprising fosmid pool sequencing and Illumina long mate-pair technologies have been used to sequence them.

Saccharum officinarum and *Saccharum spontaneum* make up the hybrid sugar cane. These two species appear strongly similar to sorghum and contain about 8–10 chromosomes. *S. officinarum* and *S. spontaneum* have a monoploid-genome, and their genome size is similar to *Sorghum bicolor* (whose genome size is 760 Mb). Backcrossing of hybrids to *S. officinarum* produces progeny with distinct chromosomal contributions from each parental species, producing commercial cultivars. Genome sequencing is complicated with the widely diverse and heterogeneous makeup of commercialized sugarcane genomes. Employing WGS techniques with sorghum genes as templates, progenitor genome sequencing might result in massive orthologous assemblies of genes (Grivet & Arruda, 2002; Jannoo et al., 2007).

Miscanthus X giganteus is an infertile triploid produced by crossing *Miscanthus sinensis* and ("tetraploid") *Miscanthus sacchariflorus. M. sinensis* had lately witnessed whole-genome duplication and solitary diploid-chromosomal blending, but none of these had ever been detected in the nearly linked sorghum genome. The wheat WGS approach might be used to measure gene copy counts and discover genetic diversity in homolog gene copies in *M. sinensis* or its hybrid and variants (Luo et al., 2009; Ming et al., 1998; Swaminathan et al., 2012).

The genomes of many agricultural plants have been decoded entirely using different sequencing methods (Table 7.1). Rice was the first crop plant to have its genome sequenced (Goff et al., 2002; Yu et al., 2002), just a few years after the initial sequencing of the model plant – *Arabidopsis thaliana* (Cao et al., 2011). The crop genome sequencing initiatives are now adapting new technology. In addition, scientists are quickly embracing second-generation sequencing strategies to sequence other desired genomes. The 430 Mbp genome of the plant *Theobroma cacao* was decoded using Roche 454 techniques (Scheffler et al., 2009). The apple genome was examined using Sanger and Roche 454 sequencing methods (4X and 12X coverage, in both) (Velasco, 2009). A similar strategy was used to create a consensus arrangement for the 504 Mbp *Vitis vinifera* genome (Velasco et al., 2007), wherein 2093 meta contigs comprising an approximated 94.6% of the genome were assembled using a mix of "6.5 X Sanger-paired-read sequences" and "4.2 X unpaired-Roche 454 reads". For the genomes of cotton plants, a hybrid of the Illumina Solexa and Roche 454 sequencing techniques were used (Wilkins et al., 2009). *Miscanthus* genome was sequenced using Roche 454 technology (Swaminathan et al., 2009). The banana genome was examined using Sanger, Illumina Solexa, and Roche 454 sequencing methods (Tuskan et al., 2006a). The 1.7 Gbps palm genome (Palmgenome, n.d.) sequencing was completed utilizing Roche 454 sequencing technology, which is a blend of WGS and BAC sequencing technology. These were the first extensive genome sequences to be finished without the need for the Sanger technique of genome sequencing.

Furthermore, the *Brassica rapa* genome was covered more than 50 times using Illumina GA II sequencing technology. The Beijing Genome Institute in Shenzhen (BGIS) and Beijing's Institute of Vegetables and Flowers (BIVF) collaborated on this study, and a high-quality genome sequence was published in early 2010. To

TABLE 7.1
Important Crop Plants for which Whole Genome Sequencing Has Been Attempted Using Numerous Sequencing Strategies

S. no	Common Name	Scientific name	Estimated Genome Size (Mb)	Sequencing Strategy	Number of Gene Predictions	Economic Importance	Year of Completion	Reference
1	Soursop	*Annona muricata*	799.11 Mb	PacBio and Illumina short-reads	23,375	Commercially grown fruit, has many medicinal applications	2021	Strijk et al. (2021)
2	Agar wood	*Aquilaria sinensis*	726.5 Mb	Illumina+nanopore+ Hi-C	29,203	Fragrant wood	2020	Ding et al. (2020)
3	Wormseed wallflower	*Erysimum cheiranthoides*	175 Mbp	PacBio and Illumina transcriptome sequencing	29,947	Model plant for studying defensive chemistry, including cardiac glycosides	2020	Züst et al. (2020)
4	Pearl millet	*Pennisetum glaucum*	1.79 Gb	WGS and bacterial artificial chromosome (BAC) sequencing	38,579	Millet species, grown as food and fodder	2017	Varshney et al. (2017)
5	Winter cress	*Barbarea vulgaris*	167.7 Mbp	Illumina GA II technology	25,350	Model plant for specialized metabolites and plant defenses	2017	Byrne et al. (2017)
6	Silver birch	*Betula pendula*	435 Mbp	454/Illumina/PacBio	28,399	Boreal forest tree, model for forest biotechnology	2017	Salojärvi et al. (2017)
7	Lavender scallops	*Kalanchoë fedtschenkoi*	256 Mbp	Illumina MiSeq platform.	30,964	Molecular genetic model for obligate CAM species in the eudicots	2017	Yang et al. (2017)

(Continued)

TABLE 7.1 *(Continued)*
Important Crop Plants for which Whole Genome Sequencing Has Been Attempted Using Numerous Sequencing Strategies

S. no	Common Name	Scientific name	Estimated Genome Size (Mb)	Sequencing Strategy	Number of Gene Predictions	Economic Importance	Year of Completion	Reference
8	Hairy bittercress	*Cardamine hirsuta*	198 Mbp	Shotgun sequencing strategy and Illumina HiSeq	29,458	A model system for studies in evolution of plant development	2016	Gan et al. (2016)
9	Commerson's nightshade	*Solanum commersonii*	838 Mbp kmer (840 Mbp)	Illumina 105x coverage	37,662	Wild potato	2015	Aversano et al. (2015)
10	Tobacco	*Nicotiana tabacum*	4,500 Mbp	Methyl filtration	90,000	Smoking	2014	Sierro et al. (2014)
11	Mung bean	*Vigna radiata*	579 Mbp	WGS	22,427	Food crop	2014	Kang et al. (2014)
12	Rape seed or oilseed rape	Brassica napus	1130 Mbp	454 GS-FLX+ Titanium and Sanger sequencing, Illumina HiSeq sequence	101,040	Oilseed crop	2014	Chalhoub et al. (2014)
13	Sugar beet	*Beta vulgaris*	714–758	WGS, Roche 454, BAC end sequencing	27,421	Sugar production	2014	Dohm et al. (2014)
14	Peach	*Prunus persica*	265 Mbp	Sanger coverage:8.47x WGS	27,852	Fruit crop	2013	Verde et al. (2013)
15	Bamboo	*Phyllostachys heterocycla*	2,075 Mbp	WGS, Illumina BAC end sequence	31,987	Building material, furniture, paper	2013	Peng et al. (2013)
16	Chickpea	*Cicer arietinum*	738 Mbp	WGS, Illumina BAC end sequence	28,269	Food crop	2013	Varshney et al. (2013)

17	Valenci, a sweet orange	*Citrus sinensis*	367 Mbp	WGS, Illumina	29,445	Fruit crop	2013	Xu et al. (2013)
18	Date palm	*Phoenix dactylifera*	671 Mbp	WGS, Illumina	41,660	Fruit crop	2011, 2013	Al-Dous et al. (2011)
19	Red rice	*Oryza rufipogon*	406 Mb	Illumina HiSeq2000 BAC physical map, Sanger sequencing	37,071	Ancestor to modern rice	2012	Huang et al. (2009)
20	Musk melon	*Cucumis melo*	450 Mbp	WGS, Roche 454, BAC end sequencing	27,427	Vegetable crop	2012	Garcia-Mas et al. (2012)
21	Banana	*Musa acuminata*	523 Mbp	WGS, Roche 454, Sanger, Illumina	36,542	Fruit crop	2012	D'Hont et al. (2012)
22	Bread wheat	*Triticum aestivum*	17,000 Mbp	WGS, Roche 454	94,000–90,000	Food crop	2012	Brenchley et al. (2012)
23	Fox millet	*Setaria italica*	490 Mbp	WGS, Sanger, Illumina, BAC end sequence	38,801	Food, fodder, and biofuel	2012	G. Zhang et al. (2012)
24	Tomato	*Solanum lycopersicum*	ca 900 Mbp	Sanger/454/Illumina/ Solid, WGS, Roche 454, BAC end Sanger sequencing	34,727	Food crop	2011, 2012	Tomato Genome Consortium (2012)
25	Neem	*Azadirachta indica*	364 Mbp	Illumina GAIIx	~20000	Source of number of terpenoids, biopesticides, and medicine	2011, 2012	Krishnan et al. (2012)

(Continued)

TABLE 7.1 *(Continued)*
Important Crop Plants for which Whole Genome Sequencing Has Been Attempted Using Numerous Sequencing Strategies

S. no	Common Name	Scientific name	Estimated Genome Size (Mb)	Sequencing Strategy	Number of Gene Predictions	Economic Importance	Year of Completion	Reference
26	Potato	*Solanum tuberosum*	726 Mbp	Sanger/454/Illumina 79.2x coverage, WGS, Roche 454	39,031	Food crop	2011	Potato Genome Sequencing Consortium et al. (2011) Hardigan et al. (2016)
27	Pigeon pea	*Cajanus cajan*	833 Mbp	WGS, Illumina	48,680	Food crop	2011	Varshney et al. (2011b)
28	Chinese cabbage	*Brassica rapa ssp. pekinensis*	485 Mbp	Illumina GA II technology, WGS, BAC end Sanger sequencing	41,174	Assorted crops and model organism	2011	X. Wang et al. (2011)
29	Hemp	*Cannabis sativa*	ca 820 Mbp	Illumina/454	30,074	Hemp and marijuana production	2011	van Bakel et al. (2011)
30	Wild strawberry	*Fragaria vesca*	240 Mbp	454/Illumina/solid 39x coverage WGS	34,809	Fruit crop	2011	Shulaev et al. (2011)
31	Soybean	*Glycine max*	1115 Mbp	Sanger sequencing and WGS	46,430	Protein and oil crop	2010	Schmutz et al. (2010)
32	Cacao	*Theobroma cacao*	430 Mbp	WGS, Sanger, Illumina, Roche 454	28,798	Food crop	2010	Argout et al. (2011)

33	Castor	*Ricinus communis*	320 Mbp	Sanger coverage ~4.6x contig	31,237	Oilseed crop	2010	J. Chen et al. (2013)
34	Maize	*Zea mays*	2.3 Gb	Sanger, 4–6x coverage per BAC, BAC physical map	39,656	Cereal crop	2009	Schnable et al. (2009b)
35	Jowar	*Sorghum bicolor*	ca. 730 Mb	Sanger, 8.5x coverage and WGS	34,496	Cereal crop	2009	Paterson et al. (2009)
36	Cucumber	*Cucumis sativus*	350 Mbp	Sanger, Illumina	26,682	Vegetable crop	2009	Huang et al. (2009)
37	Papaya	*Carica papaya*	372 Mbp	WGS	24,746	Fruit crop	2008	Ming et al. (1998)
38	Grapes	*Vitis vinifera*	475 Mbp	WGS, Sanger sequencing	30,434	Fruit crop	2007	Paterson et al. (2009)
39	Poplar	*Populus trichocarpa*	510 Mbp	WGS, BAC physical map, Sanger sequencing	73,013	Carbon sequestration, model tree, timber	2006	Tuskan et al. (2006b)

sequence the cucumber genome, researchers employed a hybrid approach that used 4X Sanger, 68X Illumina Solexa, along with 20X or 10X coverage from end-sequenced-fosmid and BAC libraries to construct greater than 96% of the genome (Huang et al., 2009). Complex BAC from barley was sequenced using Roche 454 technologies (Wicker et al., 2006; Stein, 2009), supported by repeat or duplication delineation using Illumina Solexa WGS dataset (Stein, 2009; Wicker et al., 2006). Wheat genome's enormous size and hexaploid structure posed a few challenges in determining its genomic sequence, but now wheat genome has also been published. Numerous worldwide studies have generated a large amount of efficient sequencing data for gene-rich regions. Currently, many studies are ongoing to decode the genomes of economically beneficial agricultural plants and to discover genes associated with specific traits. All identified genomes are being used in NGS technologies. In the future, genomic methods and technological advancements will keep improving the process of selection of superior genotypes, which will be a significant component in producing adequate food for the increasing population.

7.3 APPLICATION OF WHOLE-GENOME SEQUENCING IN BOTH MODEL PLANTS AND CROP PLANTS

Whole-genome sequencing (WGS) is a comprehensive method extensively used for analyzing the entire genome of an organism (Figure 7.1). WGS is widely used to analyze the genome of economically important crops to discover suitable genes that code for specific features or traits.

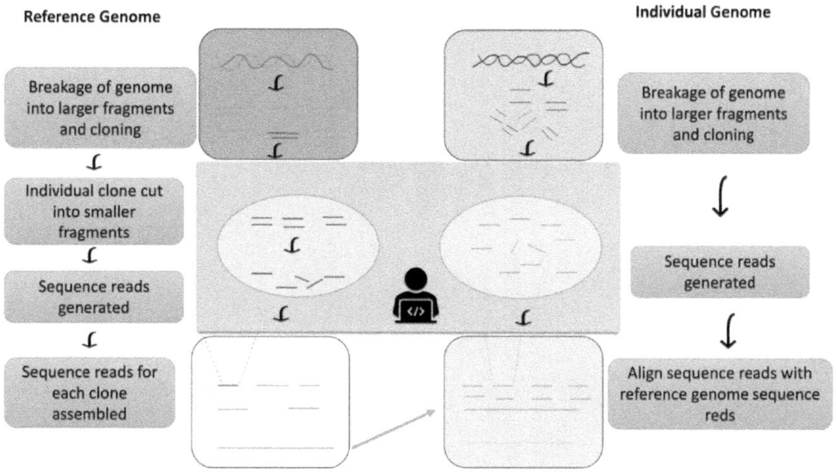

FIGURE 7.1 Schematic representation of whole-genome sequencing.

7.3.1 WGS Identified a Potential Salinity Tolerance Gene in Wild Soybean

Germplasm resources generated by whole-genome sequencing are highly beneficial for crop improvement. Essentially, it aids in investigating or identifying wild accessions that may retain a vital gene pool, which was lost during agricultural domestication. A draft genome for the wild soybean was decoded and assembled (Qi et al., 2014). After that, using genotyping-by-sequencing and phenotypic studies, a recombinant inbreeding line was created to find several QTLs (quantitative trait loci) associated with characteristics of interest. GmCHX1, a potential ion transport gene, was discovered using a blend of *de novo* sequencing datasets, and its sequence variations were linked to the salt tolerance gene. GmCHX1 has been proven to mitigate salt stress using "rapid gain of function testing" (Qi et al., 2014). Methods such as high-density marker QTL mapping by re-sequencing, whole-genome de-novo sequencing, and functional analysis can uncover genetic variation in wild soybean and aid crop improvement.

7.3.2 *Oryza brachyantha* Plant Genomic Evolution Pathways Were Discovered by Whole-Genome Sequencing

The native genera *Oryza* has a substantial pool of agronomically significant rice improvement genes that have mostly remained unexplored. *Oryza brachyantha*'s *de novo* assembled genome, with a genomic size of 261 Mb, was sequenced. The *O. brachyantha* genome is condensed with minimal retrotransposon activity and intricate deletions. Compared to the most popular crop plant, rice, the *O. brachyantha* genome contains around 32,038 genes coding for protein, of which only 70% are placed in collinear locations (J. Chen et al., 2013). The discovery of non-collinear genes proved that double-strand break repair (DSBR) via non-homologous end-joining (NHEJ) plays a key function in genetic mobility and collinearity degradation in the rice genome. In the rice genome, segmental and tandem repeats speed up the transition from euchromatin to heterochromatin, further accelerated by transposon insertions. *O. brachyantha*'s spectacular genomic sequence is a precious asset for structural, functional, and genetic investigations in the genus *Oryza* (J. Chen et al., 2013).

7.3.3 WGS Identified EMS-Induced Causative Mutations in a Non-Reference *Arabidopsis thaliana* Accession

The map-based cloning method has been the most widely utilized approach for detecting abnormalities generated by the mutagen EMS (ethyl-methane-sulfonate) in plant *Arabidopsis thaliana* (Uchida et al., 2011). However, this procedure is often labor-intensive and time-consuming. When numerous loci are associated with phenotypes, this procedure becomes more difficult. Deep sequencing methods, which require less workforce than the traditional approach of map-based cloning, may identify mutations generated by mutagen EMS. One cycle of WGS was used to quickly identify EMS-triggered mutations in the non-reference accession background whose whole genomic sequence was not publicly accessible. The causative and recombinant reporter loci were both tracked simultaneously, indicating that this strategy may be utilized to study even complicated features and patterns.

7.3.4 Investigation of *Panax notoginseng*, a Chinese Medicinal Plant

Panax notoginseng, often known as sanqi or tianqi, is a member of the Araliaceae family. It is a slow-growing plant that was documented in old Chinese medicinal literature for its capacity to promote the flow of blood and hemostasis in the body (T. Wang et al., 2016). After years of scientific research, *P. notoginseng* unique bioactive compounds, such as ginsenosides, notoginsenosides, and gypenosides, have been extracted, identified, and used for therapeutic purposes. As a medicinally important plant, generating a guide *P. notoginseng* genome is necessary to discover novel psychoactive chemicals and to demarcate their metabolic pathways. WGS technology was used to sequence, construct, and annotate the highly repetitive and complicated *P. notoginseng* genome. The genomic study of the *IPP/DMAPP* biosynthesis pathways and *TPS* genes was given using RNA sequence datasets from six tissues of *P. notoginseng*. Together with anticipated glycosyl transferase genes, this information aided in investigating recognized terpenoids in *P. notoginseng* and provided a vast gene pool for discovering potential therapeutic candidates in *Panax* species that are closely linked (T. Wang et al., 2016).

7.3.5 WGS of Multiple *Arabidopsis thaliana* Populations

Arabidopsis thaliana is found in a variety of locations across Eurasia. It is a plant with a wide range of habitats. The goal of genome sequencing was to find the genes or genetic variations that conferred environmental resistance or tolerance to this plant. In the first stage, 80 strains were collected from eight regional areas across the species' native range and sequenced on a population scale. The consequences of frequent small-scale polymorphisms and numerous more prominent insertions and deletions in the genome, gene function effect, and local and global linkage dynamics among these variations were studied. By comparing the spectra of mutations built since *A. thaliana* separated from its closest relative 10 million years ago, the functional relevance of various pathways was realized (Cao et al., 2011).

7.4 GWAS: Genome-Wide Association Studies

GWAS is an effective method that links individual genetic differences to economically relevant features and resistance/prevention from biotic and abiotic factors. This strategy entails screening organism genomes for genetic markers that may

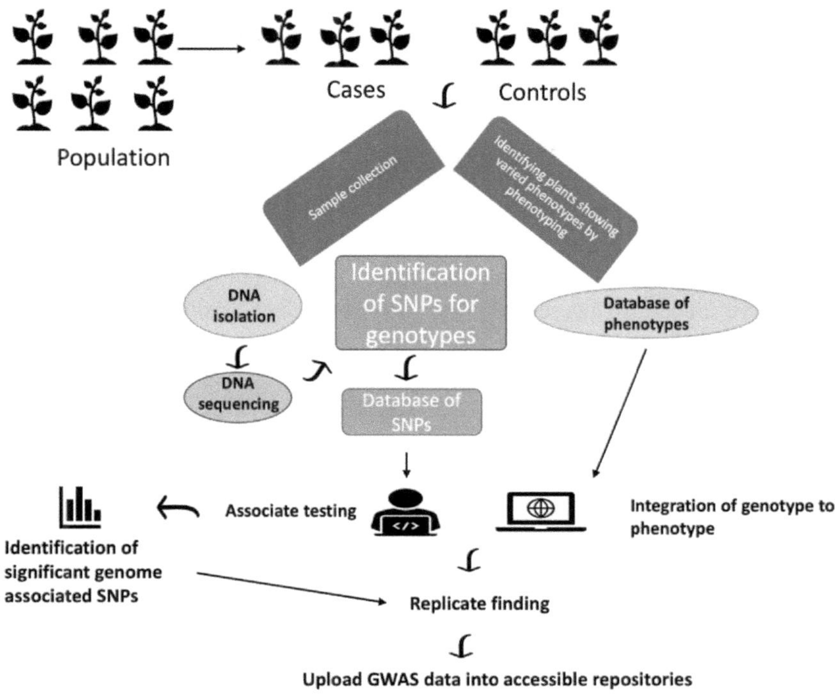

FIGURE 7.2 Schematic representation of GWAS.

be used to anticipate disease occurrence. It is a successful approach for examining many complex attributes linked to single or multiple stressors. Novel candidate genes, alleles, or QTLs essential for biotic and abiotic stress have been delineated by GWAS in numerous plants/crops (Xiao et al., 2017). GWAS is a popular method to interpret genotypic–phenotypic relationships in numerous species (Figure 7.2). Throughout the last century, over 1000 similar research have shown significant genotypic–phenotypic relationships in crop plants and given unrivalled opportunity to investigate functional genomics (H.-J. Liu & Yan, 2019). The overall purpose of GWAS is to use the most appropriate statistical approach in a specific population to relate genotypic variants to phenotypic variability. A well-known example is maize, wherein GWAS has been employed. The application of GWAS in maize has contributed to our understanding of quantitative features in the crop and advanced the breeding programs.

7.4.1 Advances and Prospects for Genome-Wide Association Studies and Genomic Selection in Crop Plant Pearl Millet

The pearl millet plant is tolerant to water shortage and resilient to climate change. It is produced in dry/arid and humid locations across the world. For the impoverished, pearl millet is a basic diet. Until the twenty-first century, it was a forgotten crop plant. Following its identification as a crop, genomic selection (GS) and GWAS were employed on pearl millet. The introduction of genotyping-by-sequencing technology (GBS), RAD-sequencing technology (RADS), and whole-genome re-sequencing technology (WGRS) made this easier. An international association mapping panel named Pearl Millet Inbred Germplasm Association Panel (PMiGAP) was created at ICRISAT in collaboration with Aberystwyth University (AU) for conducting GWAS and GS. This panel was again sequenced using the WGRS technique and included breeding lines, germplasm lines, and landraces from around 27 nations. It contains a comprehensive genome of over 29 million samples. Traits or characters similar to drought tolerance, zinc and iron content of the grain, inorganic nitrogen utilization efficacy, endosperm's starchy constituents, grain's yield components, and other features have all been mapped utilizing PMiGAP. WGRS, RAD, and tunable-genotyping-by-sequencing (t-GBS) techniques for each PMiGAP and hybrid parental lines have successfully sped up pearl millet genome screening. Using multi-environment phenotyping of numerous population groups, early efforts have been undertaken to create genomic selection models (Srivastava et al., 2019).

7.4.2 Genome-Wide Association Research (GWAR) Utilizing Whole Genome Sequencing (WGS) Quickly Uncovers Novel Genes that Influence Agronomic Traits in Rice

The GWAS is also a powerful technique for finding genes linked to economically important traits in crop species, but it is frequently hampered by population dynamics and a substantial amount of linkage disequilibrium. GWAS has been used to identify agronomically relevant genes in rice, followed by testing candidate/identified genes based upon calculated effects/results of nucleotide polymorphism. Using the same method, four additional genes linked to agronomic characters have been discovered in rice. Standard SNP analysis failed to discover some genes, but gene-based association analysis (GBAS) revealed them. Quick detection of genes related to agronomic characteristics was accomplished utilizing GWAS. This will also help to select future initiatives toward enhancement in crop production (Yano et al., 2016).

7.4.3 Application of GWAS in Plants for Abiotic Stress Tolerance

Both abiotic and biotic factors have an impact on agronomic productivity. Studies often focus on a particular gene, trait, or character when numerous factors regulate plant stress. A genome-wide association study (GWAS) is a valuable method for examining several complicated characters linked to a single or to multiple stresses. Novel gene candidates, genes, and quantitative trait loci (QTLs) relevant for biotic and abiotic stress have been deciphered by GWAS on diverse crop plants. Identifying novel genes and QTLs will help to select and develop future climate-resilient crop plants (Challa & Neelapu, 2018).

7.4.4 Application of GWAS in Plants for Biotic Stress Tolerance

The fungus *Sphacelotheca reiliana* causes head smut disease, a catastrophic worldwide infection in maize, which causes significant losses in quality and production each year. The Illumina MaizeSNP50 array was used in GWAS of head smut tolerance. A poll of 144 inbred lines yielded 45,868 single nucleotide polymorphisms, 18 linked to head smut tolerance in maize. These candidate genes were divided into three gene categories: disease response, resistance response, and additional genes that might be involved in plant resistance to disease. GWAS has proven to be an effective method for finding causative genetic variables in maize head smut resistance (K. Wang et al., 2012).

7.5 NEXT-GENERATION SEQUENCING: BRINGING A NEW ERA TO CROP RESEARCH

NGS has been critical in the advancement of genomics techniques. The diversity- and variation-related studies among people in a population are being transformed by next-generation sequencing (NGS) or second-generation sequencing technology. Compared to Sanger sequencing, which are first-generation sequencing technologies, most of these technologies lower the expense and duration necessary for DNA sequencing. The Roche/454 FLX, the Illumina/Solexa Genome Analyzer, the Applied Biosystems SOLiD System, the Helicos single-molecule sequencing, and the Pacific Biosciences SMRT equipment are currently accessible NGS technologies (Figure 7.3). These methods have allowed researchers to conduct reliable investigations using entire genomes rather than small sequences of a single gene (Gilbert, 1981).

The Roche/454 FLX technique uses pyrophosphate chemistry and synthesis-based sequencing. It was the first commercially available next-generation sequencing platform (Margulies et al., 2005). The Solexa sequencing analyzer used

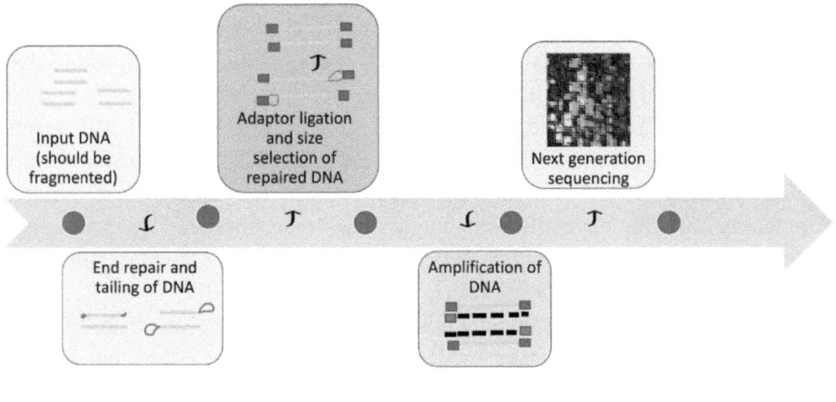

FIGURE 7.3 Schematic representation of next-generation sequencing.

sequencing-by-synthesis chemistry and was first commercialized in 2006. The SOLiD system from Life Technologies uses ligation-based sequencing (Shendure et al., 2005). Helicos tSMS technology is a brand-new technique that outperforms both existing as well as NGS technology. Helicos is the world's first platform for gene analysis that eliminates the need for amplification (Mardis, 2008; Metzker, 2010; Shendure & Ji, 2008; Zhou et al., 2010). Third-generation sequencing is a term used to describe Helicos single-molecule sequencing.

The discovery and utilization of genetic variation is the foundation of plant breeding and crop enhancement programs. Conventional phenotype-based selection is time-consuming. Genetic markers help to correlate genetic constitution and the phenotype. Several DNA-based molecular markers pertaining to key crop plants have been deciphered in recent decades and are used to identify genetic heterogeneity among cultivars (Varshney et al., 2009). The marker-assisted selection was used in the progeny, which empowers the selection of favored progeny at a very early stage. Several DNA markers have been discovered and are being utilized for plant breeding. Restriction fragment length polymorphism (RFLP), random amplification of polymorphic DNA (RAPD), amplified fragment length polymorphism (AFLP), single sequence repeat (SSR), and single nucleotide polymorphisms (SNPs) Salgotra et al. (2014) are classical markers, and the majority of them are non-selective since these are placed in non-regulatory and non-coding locations. Because of genetic recombination, there is a risk of false positives when these are utilized for marker-assisted selection (Salgotra et al., 2014). *In planta* breeding markers that recognize the allelic variation of a distinguishing characteristic are more valuable (Salgotra et al., 2014).

Next-generation sequencing (NGS) technology has recently advanced genome sequencing, allowing for the development of large numbers of new markers along with the discovery of economically important genes (Edwards & Batley, 2010). With the progress of sequencing technology, SNPs have exceeded the use of several molecular markers. Sequencing is generally done after PCR amplification for the specific genomic fragment. After that, these are aligned to find polymorphisms (Edwards & Batley, 2010). The method is both costly and time-taking. SNP discovery has become inexpensive and more efficient because of the large number of sequences produced by NGS platforms and the advent of *in silico* techniques. This method detects functional indels (insertions or deletions) within SSRs, such as partial or total gene deletions and varying repeat motifs (Salgotra et al., 2014). These markers have been used to create molecular genetics and physical maps, as well as to find genes or QTL that control economically essential features (Varshney et al., 2009). Recent advances in NGS have allowed for creating *high-density genomic maps*. NGS technologies have yielded huge volumes of sequencing data that have allowed genetic maps to be improved by increasing the density of markers. It aids in identifying related scaffolds on the map, allowing for comprehensive genetic mapping (Pérez-de-Castro et al., 2012).

Genome and transcriptome sequencing data and their expression profiles linked to various physiological situations will aid in identifying the genes that determine various traits. Detecting the fundamental genes governing a trait enables genetic manipulation to transfer the trait to other species or cultivars; so that, the desired characters can be included in a cultivated variety through marker-assisted selection (Edwards & Batley, 2010).

The use of next-generation sequencing (NGS) technologies has resulted in significant advancements in WGS, which delivers ultra-high-throughput sequences transforming plant genotyping. Genotyping-by-sequencing (GBS) is a technique that has been created and employed in the multiple samples sequencing that associates molecular marker identification with genotyping. Likewise, it has been applied to crops with large genome sizes, such as wheat and maize. GBS is a new way of detecting and genotyping SNPs in plant genomes and populations using NGS methods. Eventually, one prospective influence of genetics on crop breeding could be the evolution of system breeding methods that associate information governing gene expression together with the regulatory pathway to anticipate as well as evaluate the benefaction of genetic and epigenetic variants to phenotypes and results obtained in the field (Bevan & Uauy, 2013).

An important example of NGS application is observed in *Vigna radiata* (L.); comparative RNA-seq revealed a complex regulatory network that confers yellow mosaic disease (YMD) resistance. Mungbean YMD is one of Asia's most lethal diseases. The molecular mechanism by which plants defend against YMD is mainly undiscovered. To gain a deeper understanding of the mungbean–YMV regulatory network, an RNA-sequencing study was done involving a resistant (PMR-1) and a sensitive (var. Pusa Vishal) *Vigna* genotype in diseased as well as control environments. A group of distinctive stress-linked EST-SSRs was discovered using RNA-sequencing data to help identify associated genes with YMD tolerance. In addition, 11 transcripts connected to protection were validated using real-time PCR. The discovered network of gene sequences has shed light on the mungbean's MYMIV infection defense mechanism, which will be very effective in managing YMD resistance in mungbean (Dasgupta et al., 2021).

7.6 BIOINFORMATICS: KEY UNDERLYING OMICS ADVANCEMENTS

Bioinformatics is a fascinating science that allows us to detect, analyze, save, and retrieve biological data. It is a multidisciplinary field of study that combines aspects of computer science, biology, and statistics to create databases and software programs for analyzing biological data, such as data coming from DNA sequencing, evolutionary genetics, transcriptomics, and proteomics. Bioinformatics has applications in various fields, including medicine, microbial genomics, and agriculture (Altman, 2003).

7.6.1 Systems Bioinformatics: Increase the Precision of Computational Diagnostics and Therapies Using Network-Based Methodologies

This is a comparatively recent method that combines systems biology and traditional bioinformatics. It combines a bottom-up strategy from systems biology with a data-driven top-down method from bioinformatics to accommodate information at various levels. The growth of systems bioinformatics has been aided by the advancement of omics technology. Various omics technologies give a wide range of data, from genomics, transcriptomics, and proteomics to epigenomics, pharmacogenomics, metagenomics, and metabolomics. It is the scaffolding that applies systems techniques to analogous data, with the resolution mark and boundary of the desired system established and the promising characteristics of the system as a whole explored instead of the sum of the attributes gathered from different components of the system. An essential approach in systems bioinformatics is building numerous networks to represent each level of the omics spectrum and their unification into a multifaceted network that transports data across layers (Oulas et al., 2019). This approach not only advances agricultural biotechnology but also improves computational therapies and diagnostics, making precision medicine a reality.

7.6.2 Bioinformatics Techniques for Next-Generation Sequencing Genomics and Post-Genomics Applications

Advancements in molecular cloning, PCR, oligonucleotide microarrays, and Sanger sequencing have contributed to our current ability to study entire organismal genomes through sequencing, annotation, and research. Several next-generation sequencing technologies have emerged in recent years. The scale and cost-effectiveness of these methods were previously inconceivable, as well as their enthusiastic acceptance by scientists and the possibility for further precision and read length improvements, implying a significant and long-term impact on genomic and post-genomic biology. Data from next-generation sequencing is managed and analyzed by creating new informatics tools capable of collecting, mapping, and interpreting large amounts of nucleotide sequence data (Horner et al., 2010).

7.6.3 Using Bioinformatics Techniques to Understand Metabolomics Datasets in a Comprehensive Way

For biological investigations, having access to metabolomics data of high quality has become a must. However, analyzing those datasets in natural settings is difficult, especially since many of the discovered metabolites are not included in databases of biological pathways. Several new tools are now accessible, ranging from statistical analyses to metabolite set enrichment analysis, circuit and network visualization, pathway prediction, biochemical databases, and text mining. Incorporating them in the form of complete and impartial analysis necessitates careful consideration of both the metabolomics dataset's disclaimers as well as the biological study design's structure and features. Special considerations must be taken into account when adapting genomics methodologies for usage in metabolomics. The programming languages R and Python facilitate the interchange of diverse technologies to make it easier to share and deploy integrated processes (Barupal et al., 2018).

7.7 BIOTECHNOLOGICAL INTERVENTIONS FOR CROP IMPROVEMENT UTILIZING GENOME INFORMATION

Recent advancements in genome sequencing techniques and bioinformatics have allowed whole-genome sequencing of many crop plants, opening the door to gather information on the functional characterization of protein-coding genes and regulatory elements such as cis-acting elements in promoter regions and other non-coding elements. The interpretation/decoding of the genome of the crop plants, functional characterization of coding genes, comparing their genomes with their wild relatives, resequencing of germplasms, and marker discoveries have led to the identification of novel QTLs, SNPs. These potential candidate genes can be applied for crop improvement. There are a few widely used methods by which crop improvement programs run: molecular breeding, overexpression of one/more desired genes by genetic transformation, gene silencing by RNAi, and genome editing.

7.7.1 Molecular Breeding

Appropriate and convenient access to whole-genome sequences, GWAS, and mapping of QTLs associated with agronomic traits helps to select superior germplasm that can be utilized for genomics-assisted accelerated breeding. Accelerated breeding speeds up the process of breeding and reduces the duration from 7–8 years to 3–4 years to release a new variety that can be utilized for crop improvement. Development of wheat lines tolerant to infection by *Puccinia graminis*, drought-tolerant maize lines, drought-tolerant chickpea lines, and rice variety with improved productivity are some examples of superior crop varieties that have been generated by genomic selection and molecular breeding (Singh et al., 2020).

7.7.2 Expression and Overexpression of Candidate Genes for Desired Phenotype

The discovery of molecular cloning, the possibility of isolation and identification of beneficial genes, and the development of plant transformation techniques have facilitated the process of crop improvement by selecting a gene and transferring it to any genome by transgenic technology (Figure 7.4A). Utilizing this approach, many genetically modified crops with better quality have been developed, such as Bt cotton and Bt brinjal harboring cry gene of *Bacillus thuringiensis*, generated for better tolerance against pest attack (Singh et al., 2020); golden rice with phytoene synthase, phytoene desaturase, lycopene-β-cyclase generated rice rich in provitamin A; *Glycine max* with *GmMYB29A2* provided resistance against *Phytophthora sojae* (Jahan et al., 2020); *Solanum lycopersicum* with *PpSnRK1α* provided salinity tolerance (W.-R. Wang et al., 2020); *Triticum aestivum* with *TaCRK2* provided resistance against *Puccnia triticin* (Gu et al., 2020); *Zea mays* with *KNR6* provided enhanced grain yield (Jia et al., 2020); canola with overexpressed tryptophan decarboxylase resulted in low levels of indole glucosinolate

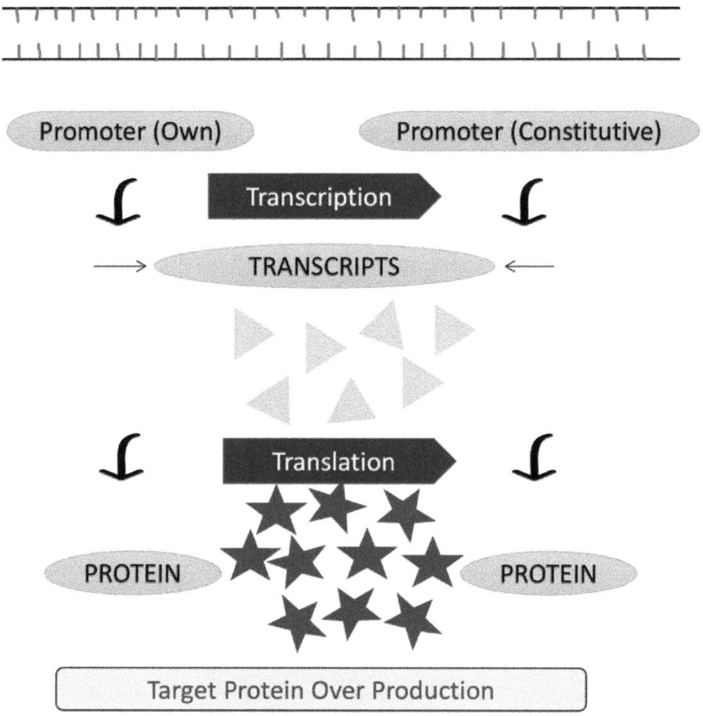

FIGURE 7.4A Over-expression of a beneficial gene through transgenic technology as a strategy for crop improvement.

(Chavadej et al., 1994); transgenic potato overexpressing amaranth albumin 1 resulted in high-protein content (Chakraborty et al., 2010) etc. This overexpression technique was also utilized to understand the function of genes to be explored for crop improvement.

7.7.3 RNA Interference for *in vivo* Knockdown of Target Genes

RNAi is an innovative technique of biotechnology that has been applied in several instances to understand the function of a gene and for genetic modification of plants by gene silencing. RNAi is associated with the generation of short interfering RNA or microRNA, which binds to the mRNA and causes its cleavage by the RISC complex, ultimately leading to the downregulation of a gene (Figure 7.4B). There are several examples of the use of RNAi for crop improvement, such as RNAi of *GmVma12* in *Glycine max* caused resistance to soybean mosaic virus (Luan et al., 2020); RNAi of *GhWRKY70D13* in cotton resulted in resistance to *Verticillium dalhiae* (Xiong et al., 2020); RNAi of OsEIL2 in *Oryza sativa* resulted in drought and salinity tolerance (Jin et al., 2020a); RNAi of *CYP74* in *Parthenium argentatum* (Placido et al., 2019) resulted in increased natural rubber

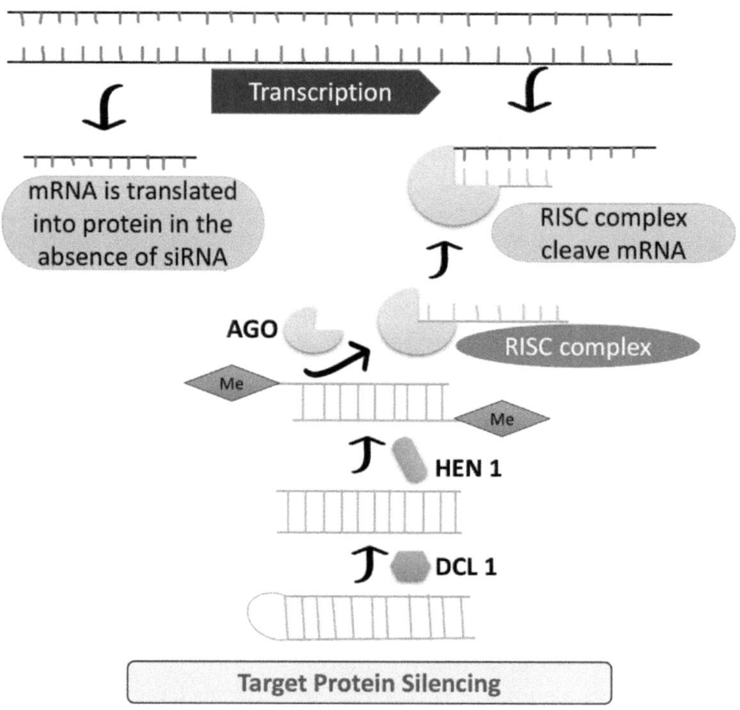

FIGURE 7.4B Schematic representation of RNA interference as a strategy for crop improvement.

production; RNAi of *syn-tasiR-TSWV* resulted in spotted wilt virus resistance (Jin et al., 2020b); RNAi of *TaDA1* in wheat caused improved kernel size (H. Liu et al., 2019); RNAi of polygalacturonase gene in tomato led to delayed fruit ripening (Sheehy et al., 1988).

7.7.4 CRISPR/CAS9-Mediated Genome Editing for Crop Improvement

So far, the gene-editing technology involving CRISPR/Cas9 has been used in more than 20 crop species for various features such as improvements in yield and environmental stress tolerance (Ricroch et al., 2017). There are innumerable reports that are treated as proof of principle studies since these reports illustrate the use of the CRISPR to knock out specialized genes involved in abiotic or biotic stress tolerance. CRISPR-mediated genome editing has improved stress tolerance such as salt stress and drought- and disease-resistance in crops to a great extent. CRISPR/Cas9 system-mediated gene editing is achieved with the help of a guide RNA (or sgRNA), and a CRISPR-linked endonuclease is two components of engineered CRISPR systems (Cas protein). The production of sgRNA and Cas9 in desired cells is enough to transform genomes, just as it is in other animal model systems. There are various vectors in plants expressing Cas9 or their variants and gRNAs, which are commercially available. To insert the desired gRNA, a scaffold sequence of gRNA and RNA polymerase III promoter (plant) on the empty gRNA backbones can be utilized (Figure 7.4C). T-DNA-mediated delivery systems are common for SSNs and gRNA. DNA-virus-mediated amplicons significantly improved gene targeting expertise (Gil-Humanes et al., 2017).

In rice, utilizing this genome editing technique to improve abiotic and biotic stress tolerance has been recently established. Examples of candidate genes that were edited using CRISPR/Cas technology are *OsGAD3* leading to GABA enriched rice; *OsALS1* for herbicide tolerance; gene editing of carotenoid cassette resulting in higher carotenoid contents etc. (Akama et al., 2020; Ding et al., 2020; Kuang et al., 2020). Wheat, an essential staple food crop all over the world (Shan et al., 2014), effectively illustrated the CRISPR/Cas9 approach to modify the TaMLO (mildew-resistant locus O) gene in wheat protoplasts. The effectiveness against powdery mildew causing *Blumeria graminis* (*Tritici*) has also been demonstrated by the CRISPR TaMLO knockdown (Y. Wang et al., 2014). Geminivirus-based DNA replicons (WDV) were employed for transitory and simple CRISPR cassettes expression in hexaploid wheat, resulting in a 12-fold increase in domestic ubiquitin gene expression. Soon, WDV-associated DNA replicons to target high-frequency genes could be a realistic technique for genetic engineering of complex genomes.

Similarly, the availability of *Gossypium hirsutum* nucleotide sequences makes it possible to perform precise changes in the DNA using CRISPR tools (Li et al., 2015). Cotton seeds contain large oil reserves, making them a vital source for the manufacture of biofuels in addition to being a fiber crop (Oliveira et al., 2016). In cotton, Janga et al. (2017) were the first to report target gene editing with the CRISPR/Cas system. In GFP-integrated transgenic cotton, three target locations in the green fluorescent protein sequence were utilized as a visible marker for phenotypic characterization. All T0 plantlets tested for gRNA2 knockdown exhibited homozygous alterations, whereas the remaining seven showed bi-allelic indels. Through homology-dependent repair, the capacity to produce DSB at the specific target position has been expanded to produce a specific nucleotide exchange or inclusion of the desired sequence of DNA (Janga et al., 2017). X. Chen et al. (2017) used two guide RNAs to test the efficacy of genome editing in cotton, one pertaining to Cloroplastos alterados 1 (GhCLA1) and the other for vacuolar H+-pyrophosphatase (GhVP). The majority of the changes in transformed plants were deletions of nucleotides, with mutational efficiency ranging from 47.6% to 81.8%. Cultivated cotton is an allotetraploid, making site-specific modifications of DNA difficult. Gao et al. (2017) and Li et al. (2017) demonstrated CRISPR/Cas9 system efficiency in producing mutations in homoeologous cotton genes and aiming several genes simultaneously in cotton using multiple sgRNAs. With the use of PCR amplification and sequencing technology in transgenic plants. C. Li et al. (2017) established CRISPR/Cas9-induced specialized trimming events in the development of cotton fiber regulating GhMYB25 homoeologous genes

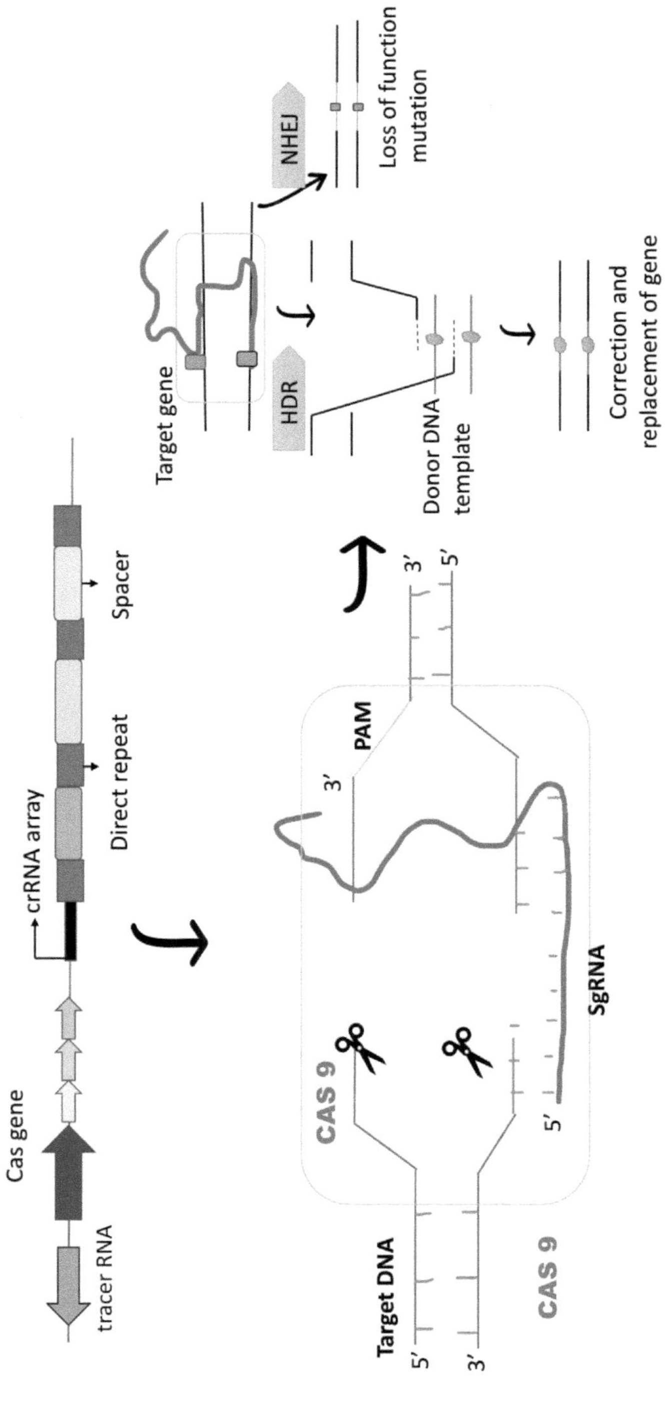

FIGURE 7.4C Schematic representation of genome editing using CRISPR/Cas 9 as a strategy for crop improvement.

(GhMYB25-like A and GhMYB25-like D). Modification of the Gh14–3–3d gene has recently been found to confer resistance to *Verticillium dahliae* infection. The disease resistance of the transgene-free plants was excellent and can serve as germplasm to develop cultivars of cotton showing disease resistance (Z. Zhang et al., 2018).

In addition to model crops, the CRISPR/Cas9 editing of the genome improved critical features in other monocots and dicots. Kapusi et al. (2017) used this technology to knock off the endo-N-acetyl-b-D-glucosaminidase (ENGase) gene in barley. The mutational efficiency of T0 and T1 mutant barley lines was 78%. Kaur et al. (2018) used CRISPR/Cas9 to modify the phytoene desaturase (RAS-PDS) gene, regulating the carotenoid biosynthetic pathway in the banana cv. Rasthali. In bananas, the use of CRISPR to knock out the RAS-PDS gene resulted in 13 mutant lines when tested for the content of chlorophyll and carotenoid concentration. This research makes way for CRISPR/Cas9 gene-editing in the banana plants, speeding up future research pertaining to incorporating desired features within the developing banana plants.

7.8 CONCLUSION

Recent advances in genome sequencing techniques, bioinformatic tools, high-throughput genotyping and phenotyping, large-scale identification of markers and functional as well as comparative genomics have generated unprecedented resources in terms of discovery of superior germplasms, identification of critical regulatory genes of a pathway, and identification of a gene/loci/marker-associated with a specific agronomic trait. These valuable resources have tremendous applications in molecular breeding and genetic engineering, such as expression of a beneficial gene, RNAi, and genome editing. These advancements in genomics and biotechnology have opened the door for sustainable agriculture.

ACKNOWLEDGMENTS

AS and IKS acknowledge SERB, India for funding. AC and AR would like to acknowledge "EVA4.0", No. CZ.02.1.01/0.0/0.0/16_019/0000803, financed by "OP RDE" and "EXTEMIT – K", No. CZ.02.1.01/0.0/0.0/15_003/0000433, financed by OP RDE and Excellent Team Grants (2023-2024), Faculty of Forestry and Wood Sciences, Czech University of Life Sciences, Prague, for financial support during the preparation of this book chapter.

REFERENCES

Akama, Y., Light, A., & Kamihira, T. (2020). *Expanding participation to design with more-than-human concerns* (p. 11). Association for Computing Machinery. https://doi.org/10.1145/3385010.3385016

Al-Dous, E. K., George, B., Al-Mahmoud, M. E., Al-Jaber, M. Y., Wang, H., Salameh, Y. M., Al-Azwani, E. K., Chaluvadi, S., Pontaroli, A. C., DeBarry, J., Arondel, V., Ohlrogge, J., Saie, I. J., Suliman-Elmeer, K. M., Bennetzen, J. L., Kruegger, R. R., &

Malek, J. A. (2011). De novo genome sequencing and comparative genomics of date palm (Phoenix dactylifera). *Nature Biotechnology*, *29*(6), 521–527. https://doi.org/10.1038/nbt.1860

Allen, A. M., Barker, G. L. A., Berry, S. T., Coghill, J. A., Gwilliam, R., Kirby, S., Robinson, P., Brenchley, R. C., D'Amore, R., McKenzie, N., Waite, D., Hall, A., Bevan, M., Hall, N., & Edwards, K. J. (2011). Transcript-specific, single-nucleotide polymorphism discovery and linkage analysis in hexaploid bread wheat (Triticum aestivum L.). *Plant Biotechnology Journal*, *9*(9), 1086–1099. https://doi.org/10.1111/j.1467-7652.2011.00628.x

Altman, R. B. (2003). The expanding scope of bioinformatics: Sequence analysis and beyond. *Heredity*, *90*(5), Article 5. https://doi.org/10.1038/sj.hdy.6800225

Argout, X., Salse, J., Aury, J.-M., Guiltinan, M. J., Droc, G., Gouzy, J., Allegre, M., Chaparro, C., Legavre, T., Maximova, S. N., Abrouk, M., Murat, F., Fouet, O., Poulain, J., Ruiz, M., Roguet, Y., Rodier-Goud, M., Barbosa-Neto, J. F., Sabot, F., . . . Lanaud, C. (2011). The genome of Theobroma cacao. *Nature Genetics*, *43*(2), 101–108. https://doi.org/10.1038/ng.736

Aversano, R., Contaldi, F., Ercolano, M. R., Grosso, V., Iorizzo, M., Tatino, F., Xumerle, L., Dal Molin, A., Avanzato, C., Ferrarini, A., Delledonne, M., Sanseverino, W., Cigliano, R. A., Capella-Gutierrez, S., Gabaldón, T., Frusciante, L., Bradeen, J. M., & Carputo, D. (2015). The Solanum commersonii genome sequence provides insights into adaptation to stress conditions and genome evolution of wild potato relatives. *The Plant Cell*, *27*(4), 954–968. https://doi.org/10.1105/tpc.114.135954

Bancroft, I., Morgan, C., Fraser, F., Higgins, J., Wells, R., Clissold, L., Baker, D., Long, Y., Meng, J., Wang, X., Liu, S., & Trick, M. (2011). Dissecting the genome of the polyploid crop oilseed rape by transcriptome sequencing. *Nature Biotechnology*, *29*(8), 762–766. https://doi.org/10.1038/nbt.1926

Barupal, D. K., Fan, S., & Fiehn, O. (2018). Integrating bioinformatics approaches for a comprehensive interpretation of metabolomics datasets. *Current Opinion in Biotechnology*, *54*, 1–9. https://doi.org/10.1016/j.copbio.2018.01.010

Batzoglou, S., Jaffe, D. B., Stanley, K., Butler, J., Gnerre, S., Mauceli, E., Berger, B., Mesirov, J. P., & Lander, E. S. (2002). ARACHNE: A whole-genome shotgun assembler. *Genome Research*, *12*(1), 177–189. https://doi.org/10.1101/gr.208902

Bevan, M. W., & Uauy, C. (2013). Genomics reveals new landscapes for crop improvement. *Genome Biology*, *14*(6), 206. https://doi.org/10.1186/gb-2013-14-6-206

Brenchley, R., Spannagl, M., Pfeifer, M., Barker, G. L. A., D'Amore, R., Allen, A. M., McKenzie, N., Kramer, M., Kerhornou, A., Bolser, D., Kay, S., Waite, D., Trick, M., Bancroft, I., Gu, Y., Huo, N., Luo, M.-C., Sehgal, S., Gill, B., . . . Hall, N. (2012). Analysis of the bread wheat genome using whole-genome shotgun sequencing. *Nature*, *491*(7426), 705–710. https://doi.org/10.1038/nature11650

Bunnik, E. M., & Le Roch, K. G. (2013). An introduction to functional genomics and systems biology. *Advances in Wound Care*, *2*(9), 490–498. https://doi.org/10.1089/wound.2012.0379

Byrne, S. L., Erthmann, P. Ø., Agerbirk, N., Bak, S., Hauser, T. P., Nagy, I., Paina, C., & Asp, T. (2017). The genome sequence of Barbarea vulgaris facilitates the study of ecological biochemistry. *Scientific Reports*, *7*, 40728. https://doi.org/10.1038/srep40728

Cao, J., Schneeberger, K., Ossowski, S., Günther, T., Bender, S., Fitz, J., Koenig, D., Lanz, C., Stegle, O., Lippert, C., Wang, X., Ott, F., Müller, J., Alonso-Blanco, C., Borgwardt, K., Schmid, K. J., & Weigel, D. (2011). Whole-genome sequencing of multiple Arabidopsis thaliana populations. *Nature Genetics*, *43*(10), 956–963. https://doi.org/10.1038/ng.911

Chakraborty, S., Chakraborty, N., Agrawal, L., Ghosh, S., Narula, K., Shekhar, S., Naik, P. S., Pande, P. C., Chakrborti, S. K., & Datta, A. (2010). Next-generation protein-rich potato expressing the seed protein gene AmA1 is a result of proteome rebalancing in transgenic tuber. *Proceedings of the National Academy of Sciences of the United States of America*, *107*(41), 17533–17538. https://doi.org/10.1073/pnas.1006265107

Chalhoub, B., Denoeud, F., Liu, S., Parkin, I. A. P., Tang, H., Wang, X., Chiquet, J., Belcram, H., Tong, C., Samans, B., Corréa, M., Da Silva, C., Just, J., Falentin, C., Koh, C. S., Le Clainche, I., Bernard, M., Bento, P., Noel, B., . . . Wincker, P. (2014). Plant genetics. Early allopolyploid evolution in the post-Neolithic Brassica napus oilseed genome. *Science (New York, N.Y.)*, *345*(6199), 950–953. https://doi.org/10.1126/science.1253435

Challa, S., & Neelapu, N. R. R. (2018). Chapter 9 – genome-wide association studies (GWAS) for abiotic stress tolerance in plants. In S. H. Wani (Ed.), *Biochemical, physiological and molecular avenues for combating abiotic stress tolerance in plants* (pp. 135–150). Academic Press. https://doi.org/10.1016/B978-0-12-813066-7.00009-7

Chavadej, S., Brisson, N., McNeil, J. N., & De Luca, V. (1994). Redirection of tryptophan leads to production of low indole glucosinolate canola. *Proceedings of the National Academy of Sciences of the United States of America*, *91*(6), 2166–2170. https://doi.org/10.1073/pnas.91.6.2166

Chen, J., Huang, Q., Gao, D., Wang, J., Lang, Y., Liu, T., Li, B., Bai, Z., Luis Goicoechea, J., Liang, C., Chen, C., Zhang, W., Sun, S., Liao, Y., Zhang, X., Yang, L., Song, C., Wang, M., Shi, J., . . . Chen, M. (2013). Whole-genome sequencing of Oryza brachyantha reveals mechanisms underlying Oryza genome evolution. *Nature Communications*, *4*, 1595. https://doi.org/10.1038/ncomms2596

Chen, X., Lu, X., Shu, N., Wang, S., Wang, J., Wang, D., Guo, L., & Ye, W. (2017). Targeted mutagenesis in cotton (Gossypium hirsutum L.) using the CRISPR/Cas9 system. *Scientific Reports*, *7*, 44304. https://doi.org/10.1038/srep44304

Dasgupta, U., Mishra, G. P., Dikshit, H. K., Mishra, D. C., Bosamia, T., Roy, A., Bhati, J., Priti, Aski, M., Kumar, R. R., & Singh, A. K. (2021). Comparative RNA-Seq analysis unfolds a complex regulatory network imparting yellow mosaic disease resistance in mungbean [*Vigna radiata* (L.) R. Wilczek]. *PLoS One*, *16*(1), e0244593.

D'Hont, A., Denoeud, F., Aury, J.-M., Baurens, F.-C., Carreel, F., Garsmeur, O., Noel, B., Bocs, S., Droc, G., Rouard, M., Da Silva, C., Jabbari, K., Cardi, C., Poulain, J., Souquet, M., Labadie, K., Jourda, C., Lengellé, J., Rodier-Goud, M., . . . Wincker, P. (2012). The banana (Musa acuminata) genome and the evolution of monocotyledonous plants. *Nature*, *488*(7410), 213–217. https://doi.org/10.1038/nature11241

Ding, X., Mei, W., Lin, Q., Wang, H., Wang, J., Peng, S., Li, H., Zhu, J., Li, W., Wang, P., Chen, H., Dong, W., Guo, D., Cai, C., Huang, S., Cui, P., & Dai, H. (2020). Genome sequence of the agarwood tree Aquilaria sinensis (Lour.) Spreng: The first chromosome-level draft genome in the Thymelaeceae family. *GigaScience*, *9*(3), giaa013. https://doi.org/10.1093/gigascience/giaa013

Dohm, J. C., Minoche, A. E., Holtgräwe, D., Capella-Gutiérrez, S., Zakrzewski, F., Tafer, H., Rupp, O., Sörensen, T. R., Stracke, R., Reinhardt, R., Goesmann, A., Kraft, T., Schulz, B., Stadler, P. F., Schmidt, T., Gabaldón, T., Lehrach, H., Weisshaar, B., & Himmelbauer, H. (2014). The genome of the recently domesticated crop plant sugar beet (Beta vulgaris). *Nature*, *505*(7484), 546–549. https://doi.org/10.1038/nature12817

Edwards, D., & Batley, J. (2010). Plant genome sequencing: Applications for crop improvement. *Plant Biotechnology Journal*, *8*(1), 2–9. https://doi.org/10.1111/j.1467-7652.2009.00459.x

Gan, X., Hay, A., Kwantes, M., Haberer, G., Hallab, A., Ioio, R. D., Hofhuis, H., Pieper, B., Cartolano, M., Neumann, U., Nikolov, L. A., Song, B., Hajheidari, M., Briskine, R., Kougioumoutzi, E., Vlad, D., Broholm, S., Hein, J., Meksem, K., . . . Tsiantis, M. (2016). The Cardamine hirsuta genome offers insight into the evolution of morphological diversity. *Nature Plants*, *2*(11), 16167. https://doi.org/10.1038/nplants.2016.167

Gao, W., Long, L., Tian, X., Xu, F., Liu, J., Singh, P. K., Botella, J. R., & Song, C. (2017). Genome editing in cotton with the CRISPR/Cas9 system. *Frontiers in Plant Science*, *8*, 1364. https://doi.org/10.3389/fpls.2017.01364

Garcia-Mas, J., Benjak, A., Sanseverino, W., Bourgeois, M., Mir, G., González, V. M., Hénaff, E., Câmara, F., Cozzuto, L., Lowy, E., Alioto, T., Capella-Gutiérrez, S., Blanca, J., Cañizares, J., Ziarsolo, P., Gonzalez-Ibeas, D., Rodríguez-Moreno, L., Droege, M., Du, L., . . . Puigdomènech, P. (2012). The genome of melon (Cucumis melo L.). *Proceedings of the National Academy of Sciences of the United States of America*, *109*(29), 11872–11877. https://doi.org/10.1073/pnas.1205415109

Gilbert, W. (1981). DNA sequencing and gene structure Nobel lecture, December 8, 1980. *Bioscience Reports*, *1*(5), 353–375. https://doi.org/10.1007/BF01116186

Gil-Humanes, J., Wang, Y., Liang, Z., Shan, Q., Ozuna, C. V., Sánchez-León, S., Baltes, N. J., Starker, C., Barro, F., Gao, C., & Voytas, D. F. (2017). High-efficiency gene targeting in hexaploid wheat using DNA replicons and CRISPR/Cas9. *The Plant Journal: For Cell and Molecular Biology*, *89*(6), 1251–1262. https://doi.org/10.1111/tpj.13446

Goff, S. A., Ricke, D., Lan, T.-H., Presting, G., Wang, R., Dunn, M., Glazebrook, J., Sessions, A., Oeller, P., Varma, H., Hadley, D., Hutchison, D., Martin, C., Katagiri, F., Lange, B. M., Moughamer, T., Xia, Y., Budworth, P., Zhong, J., . . . Briggs, S. (2002). A draft sequence of the rice genome (Oryza sativa L. ssp. Japonica). *Science (New York, N.Y.)*, *296*(5565), 92–100. https://doi.org/10.1126/science.1068275

Grivet, L., & Arruda, P. (2002). Sugarcane genomics: Depicting the complex genome of an important tropical crop. *Current Opinion in Plant Biology*, *5*(2), 122–127. https://doi.org/10.1016/s1369-5266(02)00234-0

Gu, J., Sun, J., Liu, N., Sun, X., Liu, C., Wu, L., Liu, G., Zeng, F., Hou, C., Han, S., Zhen, W., & Wang, D. (2020). A novel cysteine-rich receptor-like kinase gene, TaCRK2, contributes to leaf rust resistance in wheat. *Molecular Plant Pathology*, *21*(5), 732–746. https://doi.org/10.1111/mpp.12929

Hardigan, M. A., Crisovan, E., Hamilton, J. P., Kim, J., Laimbeer, P., Leisner, C. P., Manrique-Carpintero, N. C., Newton, L., Pham, G. M., Vaillancourt, B., Yang, X., Zeng, Z., Douches, D. S., Jiang, J., Veilleux, R. E., & Buell, C. R. (2016). Genome reduction uncovers a large dispensable genome and adaptive role for copy number variation in asexually propagated solanum tuberosum. *The Plant Cell*, *28*(2), 388–405. https://doi.org/10.1105/tpc.15.00538

Hood, L., Heath, J. R., Phelps, M. E., & Lin, B. (2004). Systems biology and new technologies enable predictive and preventative medicine. *Science (New York, NY)*, *306*(5696), 640–643. https://doi.org/10.1126/science.1104635

Horner, D. S., Pavesi, G., Castrignanò, T., De Meo, P. D., Liuni, S., Sammeth, M., Picardi, E., & Pesole, G. (2010). Bioinformatics approaches for genomics and post genomics applications of next-generation sequencing. *Briefings in Bioinformatics*, *11*(2), 181–197. https://doi.org/10.1093/bib/bbp046

Huang, S., Li, R., Zhang, Z., Li, L., Gu, X., Fan, W., Lucas, W. J., Wang, X., Xie, B., Ni, P., Ren, Y., Zhu, H., Li, J., Lin, K., Jin, W., Fei, Z., Li, G., Staub, J., Kilian, A., . . . Li, S. (2009). The genome of the cucumber, Cucumis sativus L. *Nature Genetics*, *41*(12), 1275–1281. https://doi.org/10.1038/ng.475

Jahan, M. A., Harris, B., Lowery, M., Infante, A. M., Percifield, R. J., & Kovinich, N. (2020). Glyceollin transcription factor GmMYB29A2 regulates soybean resistance to phytophthora sojae. *Plant Physiology*, *183*(2), 530–546. https://doi.org/10.1104/pp.19.01293

Janga, M. R., Campbell, L. M., & Rathore, K. S. (2017). CRISPR/Cas9-mediated targeted mutagenesis in upland cotton (Gossypium hirsutum L.). *Plant Molecular Biology*, *94*(4–5), 349–360. https://doi.org/10.1007/s11103-017-0599-3

Jannoo, N., Grivet, L., Chantret, N., Garsmeur, O., Glaszmann, J. C., Arruda, P., & D'Hont, A. (2007). Orthologous comparison in a gene-rich region among grasses reveals stability in the sugarcane polyploid genome. *The Plant Journal: For Cell and Molecular Biology*, *50*(4), 574–585. https://doi.org/10.1111/j.1365-313X.2007.03082.x

Jia, H., Li, M., Li, W., Liu, L., Jian, Y., Yang, Z., Shen, X., Ning, Q., Du, Y., Zhao, R., Jackson, D., Yang, X., & Zhang, Z. (2020). A serine/threonine protein kinase encoding gene kernel number per row 6 regulates maize grain yield. *Nature Communications*, *11*(1), 988. https://doi.org/10.1038/s41467-020-14746-7

Jin, J., Duan, J., Shan, C., Mei, Z., Chen, H., Feng, H., Zhu, J., & Cai, W. (2020a). Ethylene insensitive3-like2 (OsEIL2) confers stress sensitivity by regulating OsBURP16, the β subunit of polygalacturonase (PG1β-like) subfamily gene in rice. *Plant Science*, *292*(2020), 110353. https://doi.org/10.1016/j.plantsci.2019.110353

Jin, J., Duan, J., Shan, C., Mei, Z., Chen, H., Feng, H., Zhu, J., & Cai, W. (2020b). Ethylene insensitive3-like2 (OsEIL2) confers stress sensitivity by regulating OsBURP16, the β subunit of polygalacturonase (PG1β-like) subfamily gene in rice. *Plant Science: An International Journal of Experimental Plant Biology*, *292*, 110353. https://doi.org/10.1016/j.plantsci.2019.110353

Kang, Y. J., Kim, S. K., Kim, M. Y., Lestari, P., Kim, K. H., Ha, B.-K., Jun, T. H., Hwang, W. J., Lee, T., Lee, J., Shim, S., Yoon, M. Y., Jang, Y. E., Han, K. S., Taeprayoon, P., Yoon, N., Somta, P., Tanya, P., Kim, K. S., . . . Lee, S.-H. (2014). Genome sequence of mungbean and insights into evolution within Vigna species. *Nature Communications*, *5*, 5443. https://doi.org/10.1038/ncomms6443

Kapusi, E., Corcuera-Gómez, M., Melnik, S., & Stoger, E. (2017). Heritable genomic fragment deletions and small indels in the putative ENGase gene induced by CRISPR/Cas9 in barley. *Frontiers in Plant Science*, *8*, 540. https://doi.org/10.3389/fpls.2017.00540

Kaur, N., Alok, A., Shivani, null, Kaur, N., Pandey, P., Awasthi, P., & Tiwari, S. (2018). CRISPR/Cas9-mediated efficient editing in phytoene desaturase (PDS) demonstrates precise manipulation in banana cv. Rasthali genome. *Functional & Integrative Genomics*, *18*(1), 89–99. https://doi.org/10.1007/s10142-017-0577-5

Krishnan, N. M., Pattnaik, S., Jain, P., Gaur, P., Choudhary, R., Vaidyanathan, S., Deepak, S., Hariharan, A. K., Krishna, P. B., Nair, J., Varghese, L., Valivarthi, N. K., Dhas, K., Ramaswamy, K., & Panda, B. (2012). A draft of the genome and four transcriptomes of a medicinal and pesticidal angiosperm Azadirachta indica. *BMC Genomics*, *13*, 464. https://doi.org/10.1186/1471-2164-13-464

Kuang, Y., Li, S., Ren, B., Yan, F., Spetz, C., Li, X., Zhou, X., & Zhou, H. (2020). Base-editing-mediated artificial evolution of OsALS1 in planta to develop novel herbicide-tolerant rice germplasms. *Molecular Plant*, *13*(4), 565–572. https://doi.org/10.1016/j.molp.2020.01.010

Li, C., Unver, T., & Zhang, B. (2017). A high-efficiency CRISPR/Cas9 system for targeted mutagenesis in Cotton (Gossypium hirsutum L.). *Scientific Reports*, *7*(1), Article 1. https://doi.org/10.1038/srep43902

Li, F., Fan, G., Lu, C., Xiao, G., Zou, C., Kohel, R. J., Ma, Z., Shang, H., Ma, X., Wu, J., Liang, X., Huang, G., Percy, R. G., Liu, K., Yang, W., Chen, W., Du, X., Shi, C., Yuan, Y., . . . Yu, S. (2015). Genome sequence of cultivated Upland cotton (Gossypium hirsutum TM-1) provides insights into genome evolution. *Nature Biotechnology*, *33*(5), 524–530. https://doi.org/10.1038/nbt.3208

Liu, H., Li, H., Hao, C., Wang, K., Wang, Y., Qin, L., An, D., Li, T., & Zhang, X. (2019). TaDA1, a conserved negative regulator of kernel size, has an additive effect with TaGW2 in common wheat (Triticum aestivum L.). *Plant Biotechnology Journal*. https://doi.org/10.1111/pbi.13298

Liu, H.-J., & Yan, J. (2019). Crop genome-wide association study: A harvest of biological relevance. *The Plant Journal: For Cell and Molecular Biology*, *97*(1), 8–18. https://doi.org/10.1111/tpj.14139

Luan, H., Liao, W., Song, Y., Niu, H., Hu, T., & Zhi, H. (2020). Transgenic plant generated by RNAi - mediated knocking down of soybean Vma12 and soybean mosaic virus resistance evaluation. *AMB Express*, *10*, 62. https://doi.org/10.1186/s13568-020-00997-6

Luo, M. C., Deal, K. R., Akhunov, E. D., Akhunova, A. R., Anderson, O. D., Anderson, J. A., Blake, N., Clegg, M. T., Coleman-Derr, D., Conley, E. J., Crossman, C. C., Dubcovsky, J., Gill, B. S., Gu, Y. Q., Hadam, J., Heo, H. Y., Huo, N., Lazo, G., Ma, Y., . . . Dvorak, J. (2009). Genome comparisons reveal a dominant mechanism of chromosome number reduction in grasses and accelerated genome evolution in Triticeae. *Proceedings of the National Academy of Sciences of the United States of America*, *106*(37), 15780–15785. https://doi.org/10.1073/pnas.0908195106

Mardis, E. R. (2008). Next-generation DNA sequencing methods. *Annual Review of Genomics and Human Genetics*, *9*, 387–402. https://doi.org/10.1146/annurev.genom.9.081307.164359

Margulies, M., Egholm, M., Altman, W. E., Attiya, S., Bader, J. S., Bemben, L. A., Berka, J., Braverman, M. S., Chen, Y.-J., Chen, Z., Dewell, S. B., Du, L., Fierro, J. M., Gomes, X. V., Godwin, B. C., He, W., Helgesen, S., Ho, C. H., Irzyk, G. P., . . . Rothberg, J. M. (2005). Genome sequencing in microfabricated high-density picolitre reactors. *Nature*, *437*(7057), 376–380. https://doi.org/10.1038/nature03959

Metzker, M. L. (2010). Sequencing technologies – the next generation. *Nature Reviews: Genetics*, *11*(1), 31–46. https://doi.org/10.1038/nrg2626

Ming, R., Liu, S. C., Lin, Y. R., da Silva, J., Wilson, W., Braga, D., van Deynze, A., Wenslaff, T. F., Wu, K. K., Moore, P. H., Burnquist, W., Sorrells, M. E., Irvine, J. E., & Paterson, A. H. (1998). Detailed alignment of saccharum and sorghum chromosomes: Comparative organization of closely related diploid and polyploid genomes. *Genetics*, *150*(4), 1663–1682. https://doi.org/10.1093/genetics/150.4.1663

Nystedt, B., Street, N. R., Wetterbom, A., Zuccolo, A., Lin, Y.-C., Scofield, D. G., Vezzi, F., Delhomme, N., Giacomello, S., Alexeyenko, A., Vicedomini, R., Sahlin, K., Sherwood, E., Elfstrand, M., Gramzow, L., Holmberg, K., Hällman, J., Keech, O., Klasson, L., . . . Jansson, S. (2013). The Norway spruce genome sequence and conifer genome evolution. *Nature*, *497*(7451), 579–584. https://doi.org/10.1038/nature12211

Oliveira, M. A. C., Duarte, J. B., Morello, C. D. L., Suassuna, N. D., & Oliveira, A. B. (2016). *Mixed inheritance in the genetic control of ramulosis (Colletotrichum gossypii var. Cephalosporioides) resistance in cotton*. www.alice.cnptia.embrapa.br/handle/doc/1065802

Oulas, A., Minadakis, G., Zachariou, M., Sokratous, K., Bourdakou, M. M., & Spyrou, G. M. (2019). Systems bioinformatics: Increasing precision of computational diagnostics and therapeutics through network-based approaches. *Briefings in Bioinformatics*, *20*(3), 806–824. https://doi.org/10.1093/bib/bbx151

Palmgenome. (n.d.). Retrieved January 7, 2023, from http://tinyurl.com/palmgenome

Paterson, A. H., Bowers, J. E., Bruggmann, R., Dubchak, I., Grimwood, J., Gundlach, H., Haberer, G., Hellsten, U., Mitros, T., Poliakov, A., Schmutz, J., Spannagl, M., Tang, H., Wang, X., Wicker, T., Bharti, A. K., Chapman, J., Feltus, F. A., Gowik, U., . . . Rokhsar, D. S. (2009). The Sorghum bicolor genome and the diversification of grasses. *Nature*, *457*(7229), 551–556. https://doi.org/10.1038/nature07723

Peng, Z., Lu, Y., Li, L., Zhao, Q., Feng, Q., Gao, Z., Lu, H., Hu, T., Yao, N., Liu, K., Li, Y., Fan, D., Guo, Y., Li, W., Lu, Y., Weng, Q., Zhou, C., Zhang, L., Huang, T., . . . Jiang, Z. (2013). The draft genome of the fast-growing non-timber forest species moso bamboo (Phyllostachys heterocycla). *Nature Genetics*, *45*(4), 456–461, 461e1–461e2. https://doi.org/10.1038/ng.2569

Pérez-de-Castro, A. M., Vilanova, S., Cañizares, J., Pascual, L., Blanca, J. M., Díez, M. J., Prohens, J., & Picó, B. (2012). Application of genomic tools in plant breeding. *Current Genomics*, *13*(3), 179–195. https://doi.org/10.2174/138920212800543084

Placido, D. F., Dong, N., Dong, C., Cruz, V. M. V., Dierig, D. A., Cahoon, R. E., Kang, B.-G., Huynh, T., Whalen, M., Ponciano, G., & McMahan, C. (2019). Downregulation of a CYP74 rubber particle protein increases natural rubber production in parthenium argentatum. *Frontiers in Plant Science*, *10*, 760. https://doi.org/10.3389/fpls.2019.00760

Potato Genome Sequencing Consortium, Xu, X., Pan, S., Cheng, S., Zhang, B., Mu, D., Ni, P., Zhang, G., Yang, S., Li, R., Wang, J., Orjeda, G., Guzman, F., Torres, M., Lozano, R., Ponce, O., Martinez, D., De la Cruz, G., Chakrabarti, S. K., . . . Visser, R. G. F. (2011). Genome sequence and analysis of the tuber crop potato. *Nature*, *475*(7355), 189–195. https://doi.org/10.1038/nature10158

Qi, X., Li, M.-W., Xie, M., Liu, X., Ni, M., Shao, G., Song, C., Kay-Yuen Yim, A., Tao, Y., Wong, F.-L., Isobe, S., Wong, C.-F., Wong, K.-S., Xu, C., Li, C., Wang, Y., Guan, R., Sun, F., Fan, G., . . . Lam, H.-M. (2014). Identification of a novel salt tolerance gene in wild soybean by whole-genome sequencing. *Nature Communications*, *5*, 4340. https://doi.org/10.1038/ncomms5340

Ricroch, A., Clairand, P., & Harwood, W. (2017). Use of CRISPR systems in plant genome editing: Toward new opportunities in agriculture. *Emerging Topics in Life Sciences*, *1*(2), 169–182. https://doi.org/10.1042/ETLS20170085

Salgotra, R. K., Gupta, B. B., & Stewart, C. N. (2014). From genomics to functional markers in the era of next-generation sequencing. *Biotechnology Letters*, *36*(3), 417–426. https://doi.org/10.1007/s10529-013-1377-1

Salojärvi, J., Smolander, O.-P., Nieminen, K., Rajaraman, S., Safronov, O., Safdari, P., Lamminmäki, A., Immanen, J., Lan, T., Tanskanen, J., Rastas, P., Amiryousefi, A., Jayaprakash, B., Kammonen, J. I., Hagqvist, R., Eswaran, G., Ahonen, V. H., Serra, J. A., Asiegbu, F. O., . . . Kangasjärvi, J. (2017). Genome sequencing and population genomic analyses provide insights into the adaptive landscape of silver birch. *Nature Genetics*, *49*(6), 904–912. https://doi.org/10.1038/ng.3862

Scheffler, B. E., Kuhn, D. N., Motamayor, J. C., & Schnell, R. J. (2009). Efforts towards sequencing the Cacao genome (Theobroma cacao). *Plant and Animal Genomes, XVII.*

Schmutz, J., Cannon, S. B., Schlueter, J., Ma, J., Mitros, T., Nelson, W., Hyten, D. L., Song, Q., Thelen, J. J., Cheng, J., Xu, D., Hellsten, U., May, G. D., Yu, Y., Sakurai, T., Umezawa, T., Bhattacharyya, M. K., Sandhu, D., Valliyodan, B., . . . Jackson, S. A. (2010). Genome sequence of the palaeopolyploid soybean. *Nature*, *463*(7278), 178–183. https://doi.org/10.1038/nature08670

Schnable, P. S., Ware, D., Fulton, R. S., Stein, J. C., Wei, F., Pasternak, S., Liang, C., Zhang, J., Fulton, L., Graves, T. A., Minx, P., Reily, A. D., Courtney, L., Kruchowski, S. S., Tomlinson, C., Strong, C., Delehaunty, K., Fronick, C., Courtney, B., . . . Wilson, R. K. (2009a). The B73 maize genome: Complexity, diversity, and dynamics. *Science (New York, NY)*, *326*(5956), 1112–1115. https://doi.org/10.1126/science.1178534

Schnable, P. S., Ware, D., Fulton, R. S., Stein, J. C., Wei, F., Pasternak, S., Liang, C., Zhang, J., Fulton, L., Graves, T. A., Minx, P., Reily, A. D., Courtney, L., Kruchowski, S. S., Tomlinson, C., Strong, C., Delehaunty, K., Fronick, C., Courtney, B., . . . Wilson, R. K. (2009b). The B73 maize genome: Complexity, diversity, and dynamics. *Science (New York, NY)*, *326*(5956), 1112–1115. https://doi.org/10.1126/science.1178534

Shan, Q., Wang, Y., Li, J., & Gao, C. (2014). Genome editing in rice and wheat using the CRISPR/Cas system. *Nature Protocols*, *9*(10), 2395–2410. https://doi.org/10.1038/nprot.2014.157

Sheehy, R. E., Kramer, M., & Hiatt, W. R. (1988). Reduction of polygalacturonase activity in tomato fruit by antisense RNA. *Proceedings of the National Academy of Sciences of the United States of America*, *85*(23), 8805–8809. https://doi.org/10.1073/pnas.85.23.8805

Shendure, J., & Ji, H. (2008). Next-generation DNA sequencing. *Nature Biotechnology*, *26*(10), 1135–1145. https://doi.org/10.1038/nbt1486

Shendure, J., Porreca, G. J., Reppas, N. B., Lin, X., McCutcheon, J. P., Rosenbaum, A. M., Wang, M. D., Zhang, K., Mitra, R. D., & Church, G. M. (2005). Accurate multiplex polony sequencing of an evolved bacterial genome. *Science (New York, NY)*, *309*(5741), 1728–1732. https://doi.org/10.1126/science.1117389

Shulaev, V., Sargent, D. J., Crowhurst, R. N., Mockler, T. C., Folkerts, O., Delcher, A. L., Jaiswal, P., Mockaitis, K., Liston, A., Mane, S. P., Burns, P., Davis, T. M., Slovin, J. P., Bassil, N., Hellens, R. P., Evans, C., Harkins, T., Kodira, C., Desany, B., . . . Folta, K. M. (2011). The genome of woodland strawberry (Fragaria vesca). *Nature Genetics*, *43*(2), 109–116. https://doi.org/10.1038/ng.740

Sierro, N., Battey, J. N. D., Ouadi, S., Bakaher, N., Bovet, L., Willig, A., Goepfert, S., Peitsch, M. C., & Ivanov, N. V. (2014). The tobacco genome sequence and its comparison with those of tomato and potato. *Nature Communications*, *5*, 3833. https://doi.org/10.1038/ncomms4833

Singh, R. K., Prasad, A., Muthamilarasan, M., Parida, S. K., & Prasad, M. (2020). Breeding and biotechnological interventions for trait improvement: Status and prospects. *Planta*, *252*(4), 54. https://doi.org/10.1007/s00425-020-03465-4

Srivastava, R. K., Singh, R. B., Pujarula, V. L., Bollam, S., Pusuluri, M., Chellapilla, T. S., Yadav, R. S., & Gupta, R. (2019). Genome-wide association studies and genomic selection in pearl millet: Advances and prospects. *Frontiers in Genetics*, *10*, 1389. https://doi.org/10.3389/fgene.2019.01389

Stein, N. (2009). Barley genome sequencing: First steps. *Plant and Animal Genomes*, *XVII*.

Strijk, J. S., Hinsinger, D. D., Roeder, M. M., Chatrou, L. W., Couvreur, T. L. P., Erkens, R. H. J., Sauquet, H., Pirie, M. D., Thomas, D. C., & Cao, K. (2021). Chromosome-level reference genome of the soursop (Annona muricata): A new resource for Magnoliid research and tropical pomology. *Molecular Ecology Resources*, *21*(5), 1608–1619. https://doi.org/10.1111/1755-0998.13353

Swaminathan, K., Chae, W. B., Mitros, T., Varala, K., Xie, L., Barling, A., Glowacka, K., Hall, M., Jezowski, S., Ming, R., Hudson, M., Juvik, J. A., Rokhsar, D. S., & Moose, S. P. (2012). A framework genetic map for Miscanthus sinensis from RNAseq-based markers shows recent tetraploidy. *BMC Genomics*, *13*, 142. https://doi.org/10.1186/1471-2164-13-142

Swaminathan, K., Varala, K., Moose, S. P., Rokhsar, D., Ming, R., & Hudson, M. E. (2009). A genome survey of Miscanthus Giganteus. *Plant and Animal Genomes*, *XVII*.

Tomato Genome Consortium. (2012). The tomato genome sequence provides insights into fleshy fruit evolution. *Nature*, *485*(7400), 635–641. https://doi.org/10.1038/nature11119

Town, C. D., Cheung, F., Maiti, R., Crabtree, J., Haas, B. J., Wortman, J. R., Hine, E. E., Althoff, R., Arbogast, T. S., Tallon, L. J., Vigouroux, M., Trick, M., & Bancroft, I. (2006). Comparative genomics of Brassica oleracea and Arabidopsis thaliana reveal gene loss, fragmentation, and dispersal after polyploidy. *The Plant Cell*, *18*(6), 1348–1359. https://doi.org/10.1105/tpc.106.041665

Tuskan, G. A., Difazio, S., Jansson, S., Bohlmann, J., Grigoriev, I., Hellsten, U., Putnam, N., Ralph, S., Rombauts, S., Salamov, A., Schein, J., Sterck, L., Aerts, A., Bhalerao, R. R., Bhalerao, R. P., Blaudez, D., Boerjan, W., Brun, A., Brunner, A., . . .

Rokhsar, D. (2006b). The genome of black cottonwood, Populus trichocarpa (Torr. & Gray). *Science (New York, NY)*, *313*(5793), 1596–1604. https://doi.org/10.1126/science.1128691

Tuskan, G. A., Difazio, S., Jansson, S., Bohlmann, J., Grigoriev, I., Hellsten, U., Putnam, N., Ralph, S., Rombauts, S., Salamov, A., Schein, J., Sterck, L., Aerts, A., Bhalerao, R. R., Bhalerao, R. P., Blaudez, D., Boerjan, W., Brun, A., Brunner, A., . . . Rokhsar, D. (2006a). The genome of black cottonwood, Populus trichocarpa (Torr. & Gray). *Science (New York, NY)*, *313*(5793), 1596–1604. https://doi.org/10.1126/science.1128691

Uchida, N., Sakamoto, T., Kurata, T., & Tasaka, M. (2011). Identification of EMS-induced causal mutations in a non-reference Arabidopsis thaliana accession by whole genome sequencing. *Plant & Cell Physiology*, *52*(4), 716–722. https://doi.org/10.1093/pcp/pcr029

van Bakel, H., Stout, J. M., Cote, A. G., Tallon, C. M., Sharpe, A. G., Hughes, T. R., & Page, J. E. (2011). The draft genome and transcriptome of Cannabis sativa. *Genome Biology*, *12*(10), R102. https://doi.org/10.1186/gb-2011-12-10-r102

Varshney, R. K., Chen, W., Li, Y., Bharti, A. K., Saxena, R. K., Schlueter, J. A., Donoghue, M. T. A., Azam, S., Fan, G., Whaley, A. M., Farmer, A. D., Sheridan, J., Iwata, A., Tuteja, R., Penmetsa, R. V., Wu, W., Upadhyaya, H. D., Yang, S.-P., Shah, T., . . . Jackson, S. A. (2011a). Draft genome sequence of pigeonpea (Cajanus cajan), an orphan legume crop of resource-poor farmers. *Nature Biotechnology*, *30*(1), 83–89. https://doi.org/10.1038/nbt.2022

Varshney, R. K., Chen, W., Li, Y., Bharti, A. K., Saxena, R. K., Schlueter, J. A., Donoghue, M. T. A., Azam, S., Fan, G., Whaley, A. M., Farmer, A. D., Sheridan, J., Iwata, A., Tuteja, R., Penmetsa, R. V., Wu, W., Upadhyaya, H. D., Yang, S.-P., Shah, T., . . . Jackson, S. A. (2011b). Draft genome sequence of pigeonpea (Cajanus cajan), an orphan legume crop of resource-poor farmers. *Nature Biotechnology*, *30*(1), 83–89. https://doi.org/10.1038/nbt.2022

Varshney, R. K., Nayak, S. N., May, G. D., & Jackson, S. A. (2009). Next-generation sequencing technologies and their implications for crop genetics and breeding. *Trends in Biotechnology*, *27*(9), 522–530. https://doi.org/10.1016/j.tibtech.2009.05.006

Varshney, R. K., Shi, C., Thudi, M., Mariac, C., Wallace, J., Qi, P., Zhang, H., Zhao, Y., Wang, X., Rathore, A., Srivastava, R. K., Chitikineni, A., Fan, G., Bajaj, P., Punnuri, S., Gupta, S. K., Wang, H., Jiang, Y., Couderc, M., . . . Xu, X. (2017). Pearl millet genome sequence provides a resource to improve agronomic traits in arid environments. *Nature Biotechnology*, *35*(10), 969–976. https://doi.org/10.1038/nbt.3943

Varshney, R. K., Song, C., Saxena, R. K., Azam, S., Yu, S., Sharpe, A. G., Cannon, S., Baek, J., Rosen, B. D., Tar'an, B., Millan, T., Zhang, X., Ramsay, L. D., Iwata, A., Wang, Y., Nelson, W., Farmer, A. D., Gaur, P. M., Soderlund, C., . . . Cook, D. R. (2013). Draft genome sequence of chickpea (Cicer arietinum) provides a resource for trait improvement. *Nature Biotechnology*, *31*(3), 240–246. https://doi.org/10.1038/nbt.2491

Velasco, R. (2009). The golden delicious apple genome: An international whole genome sequencing initiative. *Plant and Animal Genomes*, *XVII*.

Velasco, R., Zharkikh, A., Affourtit, J., Dhingra, A., Cestaro, A., Kalyanaraman, A., Fontana, P., Bhatnagar, S. K., Troggio, M., Pruss, D., Salvi, S., Pindo, M., Baldi, P., Castelletti, S., Cavaiuolo, M., Coppola, G., Costa, F., Cova, V., Dal Ri, A., . . . Viola, R. (2010). The genome of the domesticated apple (Malus × domestica Borkh.). *Nature Genetics*, *42*(10), 833–839. https://doi.org/10.1038/ng.654

Velasco, R., Zharkikh, A., Troggio, M., Cartwright, D. A., Cestaro, A., Pruss, D., Pindo, M., Fitzgerald, L. M., Vezzulli, S., Reid, J., Malacarne, G., Iliev, D., Coppola, G., Wardell, B., Micheletti, D., Macalma, T., Facci, M., Mitchell, J. T., Perazzolli, M., . . . Viola, R. (2007). A high quality draft consensus sequence of the genome of a heterozygous grapevine variety. *PloS One*, *2*(12), e1326. https://doi.org/10.1371/journal.pone.0001326

Verde, I., Abbott, A. G., Scalabrin, S., Jung, S., Shu, S., Marroni, F., Zhebentyayeva, T., Dettori, M. T., Grimwood, J., Cattonaro, F., Zuccolo, A., Rossini, L., Jenkins, J., Vendramin, E., Meisel, L. A., Decroocq, V., Sosinski, B., Prochnik, S., Mitros, T., . . . Rokhsar, D. S. (2013). The high-quality draft genome of peach (Prunus persica) identifies unique patterns of genetic diversity, domestication and genome evolution. *Nature Genetics*, *45*(5), Article 5. https://doi.org/10.1038/ng.2586

Wang, K., Wang, Z., Li, F., Ye, W., Wang, J., Song, G., Yue, Z., Cong, L., Shang, H., Zhu, S., Zou, C., Li, Q., Yuan, Y., Lu, C., Wei, H., Gou, C., Zheng, Z., Yin, Y., Zhang, X., . . . Yu, S. (2012). The draft genome of a diploid cotton Gossypium raimondii. *Nature Genetics*, *44*(10), 1098–1103. https://doi.org/10.1038/ng.2371

Wang, M., Yan, J., Zhao, J., Song, W., Zhang, X., Xiao, Y., & Zheng, Y. (2012). Genome-wide association study (GWAS) of resistance to head smut in maize. *Plant Science: An International Journal of Experimental Plant Biology*, *196*, 125–131. https://doi.org/10.1016/j.plantsci.2012.08.004

Wang, T., Guo, R., Zhou, G., Zhou, X., Kou, Z., Sui, F., Li, C., Tang, L., & Wang, Z. (2016). Traditional uses, botany, phytochemistry, pharmacology and toxicology of Panax notoginseng (Burk.) F. H. Chen: A review. *Journal of Ethnopharmacology*, *188*, 234–258. https://doi.org/10.1016/j.jep.2016.05.005

Wang, W.-R., Liang, J.-H., Wang, G.-F., Sun, M.-X., Peng, F.-T., & Xiao, Y.-S. (2020). Overexpression of PpSnRK1α in tomato enhanced salt tolerance by regulating ABA signaling pathway and reactive oxygen metabolism. *BMC Plant Biology*, *20*(1), 128. https://doi.org/10.1186/s12870-020-02342-2

Wang, X., Wang, H., Wang, J., Sun, R., Wu, J., Liu, S., Bai, Y., Mun, J.-H., Bancroft, I., Cheng, F., Huang, S., Li, X., Hua, W., Wang, J., Wang, X., Freeling, M., Pires, J. C., Paterson, A. H., Chalhoub, B., . . . Brassica rapa Genome Sequencing Project Consortium. (2011). The genome of the mesopolyploid crop species Brassica rapa. *Nature Genetics*, *43*(10), 1035–1039. https://doi.org/10.1038/ng.919

Wang, Y., Cheng, X., Shan, Q., Zhang, Y., Liu, J., Gao, C., & Qiu, J.-L. (2014). Simultaneous editing of three homoeoalleles in hexaploid bread wheat confers heritable resistance to powdery mildew. *Nature Biotechnology*, *32*(9), 947–951. https://doi.org/10.1038/nbt.2969

Wicker, T., Schlagenhauf, E., Graner, A., Close, T. J., Keller, B., & Stein, N. (2006). 454 sequencing put to the test using the complex genome of barley. *BMC Genomics*, *7*, 275. https://doi.org/10.1186/1471-2164-7-275

Wilkins, T. A., Mudge, J., Abidi, N., Allen, R., & Auld, D. (2009). The sequencing and resequencing of cotton. *Plant and Animal Genomes, XVII*.

Xiao, Y., Liu, H., Wu, L., Warburton, M., & Yan, J. (2017). Genome-wide association studies in maize: Praise and stargaze. *Molecular Plant*, *10*(3), 359–374. https://doi.org/10.1016/j.molp.2016.12.008

Xiong, X. P., Sun, S. C., Zhang, X. Y., Li, Y. J., Liu, F., Zhu, Q. H., Xue, F., & Sun, J. (2020, February). GhWRKY70D13 regulates resistance to Verticillium dahliae in cotton through the ethylene and jasmonic acid signaling pathways. *Frontiers in Plant Science*, *11*, 00069. https://doi.org/10.3389/fpls.2020.00069

Xu, Q., Chen, L.-L., Ruan, X., Chen, D., Zhu, A., Chen, C., Bertrand, D., Jiao, W.-B., Hao, B.-H., Lyon, M. P., Chen, J., Gao, S., Xing, F., Lan, H., Chang, J.-W., Ge, X., Lei, Y., Hu, Q., Miao, Y., . . . Ruan, Y. (2013). The draft genome of sweet orange (Citrus sinensis). *Nature Genetics*, *45*(1), 59–66. https://doi.org/10.1038/ng.2472

Yang, X., Hu, R., Yin, H., Jenkins, J., Shu, S., Tang, H., Liu, D., Weighill, D. A., Cheol Yim, W., Ha, J., Heyduk, K., Goodstein, D. M., Guo, H.-B., Moseley, R. C., Fitzek, E., Jawdy, S., Zhang, Z., Xie, M., Hartwell, J., . . . Tuskan, G. A. (2017). The Kalanchoë genome provides insights into convergent evolution and building blocks of crassulacean acid metabolism. *Nature Communications*, *8*(1), 1899. https://doi.org/10.1038/s41467-017-01491-7

Yano, K., Yamamoto, E., Aya, K., Takeuchi, H., Lo, P.-C., Hu, L., Yamasaki, M., Yoshida, S., Kitano, H., Hirano, K., & Matsuoka, M. (2016). Genome-wide association study using whole-genome sequencing rapidly identifies new genes influencing agronomic traits in rice. *Nature Genetics*, *48*(8), 927–934. https://doi.org/10.1038/ng.3596

Young, N. D., Debellé, F., Oldroyd, G. E. D., Geurts, R., Cannon, S. B., Udvardi, M. K., Benedito, V. A., Mayer, K. F. X., Gouzy, J., Schoof, H., Van de Peer, Y., Proost, S., Cook, D. R., Meyers, B. C., Spannagl, M., Cheung, F., De Mita, S., Krishnakumar, V., Gundlach, H., . . . Roe, B. A. (2011). The Medicago genome provides insight into the evolution of rhizobial symbioses. *Nature*, *480*(7378), 520–524. https://doi.org/10.1038/nature10625

Yu, J., Hu, S., Wang, J., Wong, G. K.-S., Li, S., Liu, B., Deng, Y., Dai, L., Zhou, Y., Zhang, X., Cao, M., Liu, J., Sun, J., Tang, J., Chen, Y., Huang, X., Lin, W., Ye, C., Tong, W., . . . Yang, H. (2002). A draft sequence of the rice genome (Oryza sativa L. ssp. Indica). *Science (New York, NY)*, *296*(5565), 79–92. https://doi.org/10.1126/science.1068037

Zhang, G., Liu, X., Quan, Z., Cheng, S., Xu, X., Pan, S., Xie, M., Zeng, P., Yue, Z., Wang, W., Tao, Y., Bian, C., Han, C., Xia, Q., Peng, X., Cao, R., Yang, X., Zhan, D., Hu, J., . . . Wang, J. (2012). Genome sequence of foxtail millet (Setaria italica) provides insights into grass evolution and biofuel potential. *Nature Biotechnology*, *30*(6), 549–554. https://doi.org/10.1038/nbt.2195

Zhang, Z., Ge, X., Luo, X., Wang, P., Fan, Q., Hu, G., Xiao, J., Li, F., & Wu, J. (2018). Simultaneous editing of two copies of Gh14–3–3d confers enhanced transgene-clean plant defense against verticillium dahliae in allotetraploid upland cotton. *Frontiers in Plant Science*, *9*. www.frontiersin.org/articles/10.3389/fpls.2018.00842

Zhou, X., Ren, L., Meng, Q., Li, Y., Yu, Y., & Yu, J. (2010). The next-generation sequencing technology and application. *Protein & Cell*, *1*(6), 520–536. https://doi.org/10.1007/s13238-010-0065-3

Züst, T., Strickler, S. R., Powell, A. F., Mabry, M. E., An, H., Mirzaei, M., York, T., Holland, C. K., Kumar, P., Erb, M., Petschenka, G., Gómez, J.-M., Perfectti, F., Müller, C., Pires, J. C., Mueller, L. A., & Jander, G. (2020). Independent evolution of ancestral and novel defenses in a genus of toxic plants (Erysimum, Brassicaceae). *ELife*, *9*, e51712. https://doi.org/10.7554/eLife.51712

8 Epigenetic Response During Plant–Pathogenic Interaction
An Overview

Chitra Nehra, Dinesh Yadav, and R. K. Gaur

8.1 INTRODUCTION

The sessile nature of plants makes them available for constant encounters with various biotic and abiotic stresses. To adapt and survive in stressed environments, plants require changes that help to mitigate short-term negative effects without affecting future fitness (Douma et al., 2017). A key mechanism adopted by plants is control of gene expression when subjected to adverse environmental conditions (Yaish et al., 2017). These changes are called epigenetic changes when there is heritable alteration in gene expression and phenotype variation without any change in DNA sequence (Wu & Morris, 2001). Epigenetic regulation offers stress tolerance and improves plants' survivability (Hirayama & Shinozaki, 2010; Richards, 2006). Epigenetic regulations create chemical changes at the molecular level that leads to altering gene expression. A series of organized epigenetic events manifested on the nucleosome level includes DNA methylation, histone post-transitional modifications and small RNA interference, revealing changes in gene expression (Akhter et al., 2021; Baulcombe & Dean, 2014; Defraia et al., 2013; Ramos-Cruz et al., 2021; Zhu et al., 2016).

These changes collectively control gene functions and regulations at a cellular level in both normal and stressed conditions (Chen et al., 2013; Vaissière et al., 2008). These modifications may be inherited through mitosis/meiosis, and it has been reported in many experiments that stress-induced changes can be transferred to the next generations (Henderson & Jacobsen, 2007; Kumar & Singh, 2016; Saze, 2008; Stassen et al., 2018; Thiebaut et al., 2019). Epigenetics is useful in plant breeding to evaluate transmission of epigenetic markers within generations to develop desired traits in crops (Slotkin, 2016; Yang et al., 2020). Epigenetic changes help in assessing evolutionary process like natural selection, inheritance etc.; however, epigenetic and phenotypic changes can vary in natural environments that make it difficult to correlate phenotypic variations with certain epigenetic changes (Laland et al., 2014; Pigliucci, 2007; Richards et al., 2010; Xia et al., 2016). Environmental factors have also been found to induce phenotypic

DOI: 10.1201/9781003153481-8

variations, so it is possible that both epigenetic modifications and environmental factors together affect plant response to unfavorable conditions. However, the exact contribution of both to create phenotypic variation is still not clear (Richards, 2011; Schmitz & Ecker, 2012; Wilschut et al., 2016; Xia et al., 2016). Epigenetics can be used as a potential tool in agriculture and crop improvement as it offers an understanding of the molecular process to create natural selection and phenotypic variations in response to environmental challenges (Dowen et al., 2012; Kumar, 2018). Epigenetic modifications are important in plant stress response mechanisms, without causing any alteration in DNA sequences. In this chapter, we have discussed and updated our understanding of epigenetic changes and their mechanisms that may help plants with biotic stress adaptations (Figure 8.1).

8.2 TYPES OF EPIGENETIC MODIFICATIONS

8.2.1 DNA Methylation

DNA methylation causes many inheritable, adaptive epigenetic modification in plants (Hirsch et al., 2013). DNA methylation is an addition of a methyl group on a fifth carbon of cytosine residue of the DNA sequence; this process is catalyzed by cytosine methyltransferase enzymes (Bewick et al., 2016; Niederhuth & Schmitz, 2017). The added methyl group alters gene expression by the attachment of different proteins, which bring changes in histone scaffolds and consequent alteration of gene expression (Klose & Bird, 2006). Methylation in promoter sequences and transposable elements causes gene silencing, while methylation within the gene induces its expression (Chan et al., 2005). Higher levels of genomic DNA methylation down-regulate gene expression, which allows the plant to save energy during stress exposure, while in resistance-related genes, less methylation enables activation of chromatin, which leads to expression of genes and developing long-term stress resistance in plants. In some plant species, DNA methylation or demethylation are induced by various environmental stresses (Al-Lawati et al., 2016; Yaish, 2013). These modifications can be transmitted in the next generations, even without the presence of the original stress conditions. Although DNA methylation varies among different plant species and stress conditions. Different epigenetic variations like hypermethylation at the genome level, hypomethylation of resistance responsive genes and methylation of biotic stress-responsive genes might help in deciphering plant response to biotic stresses (Bender, 2004; Dowen et al., 2012).

In plants, methylation is found most commonly in gene promoter regions and both heterochromatins as well euchromatin regions (Bird, 2002; Ficz et al., 2013). Extremely methylated sequences and transposable elements occur at heterochromatin regions while at euchromatin regions lesser 5-mc is present (Akimoto et al., 2007; Richards, 2006). Methylation is categorized into symmetrical or asymmetrical methylation-based target sequences; CG and CHG (H = A, C or T) methylation are referred to as symmetrical, while CHH methylation is asymmetrical.

In plants, methylation is found most commonly at evolutionary conserved CpG sites (Lennartsson, 2017). Symmetric sites present in CpG islands containing methyl-able sets of cytosines on separate strands, while asymmetric methylation sites have cytosine in any sequence (Agius et al., 2006; Grant-Downton & Dickinson, 2005; Piccolo & Fisher, 2014). In the *Arabidopsis* genome, methylomic and transcriptomic studies have demonstrated different DNA methylation patterns and their possible effects on gene expression and physiological characteristics (Paran & van der Knaap, 2007).

8.2.1.1 Mechanism of DNA Methylation

Symmetrical methylation at CpG and CpNpG helps in replication, while in non-symmetric methylation of CpNpN, an essential demethylation step occurs during replication (Pikaard et al., 2008). Mostly methylation is found in CpG sites; only some occur at non-CpG sites (Bostick et al., 2007). During DNA replication, an unmethylated, newly synthesized strand makes a hemimethylated medium that helps in recruiting methyltransferase 1 followed by methylation of other unmethylated CG sites (Gouil & Baulcombe, 2016). CHH methylation happens through two separate mechanisms. In heterochromatin regions, chromomethylase 2 enzymes produce H3K9me2 in transposable elements (Stroud et al., 2014; Zhao & Chen, 2014). In euchromatin regions of DNA sequences, the siRNAs mediated de novo methyltransferase creates a loop that directs methylation process (Niehrs, 2009; Paran & van der Knaap, 2007). In plants, DNA methylation is mediated by different DNA methyltransferase enzymes according to sequences, and after establishment it is upheld by methyltransferase 1 and chromomethylase 3 (Li et al., 2020; Simon & Meyers, 2011). In the active DNA methylation process, either oxidation or deamination of methylated nitrogenous base takes place. This edited nucleotide with other neighboring nucleotides is replaced by base excision pathways (Bochtler et al., 2017).

8.2.1.2 DNA Methylation and Biotic Stresses

In response to stresses, DNA methylation can alter dynamically (Zhang et al., 2018) and a methylation pattern can be transferred from mother to daughter cells to inherit epigenetic modifications. In plant defense, DNA methylation plays an important role (Hewezi et al., 2018; Ramirez-Prado et al., 2018), and in most studies it has been found that resistance is enhanced when there is loss of methylation, which means plant defense is negatively regulated by DNA methylation (López Sánchez et al., 2016). This has been proved in *Arabidopsis* experiments, where DNA methylation depleted mutants like triple mutant drm1 drm2 drm3 or met 1 met 1–3 showed induced defense-related gene expression and more resistance to harmful bacteria *Pseudomonas syringae* (Pst) than wild-type Columbia-0 plants (Dowen et al., 2012; Espinas et al., 2016). Although methylation patterns were different in these mutants, drm1 drm2 drm3 has CHH methylation and depleted CHG methylation, whereas met 1–3 has less global CG methylation (Stroud et al., 2013). These different methylation patterns suggest the possibility of a center at

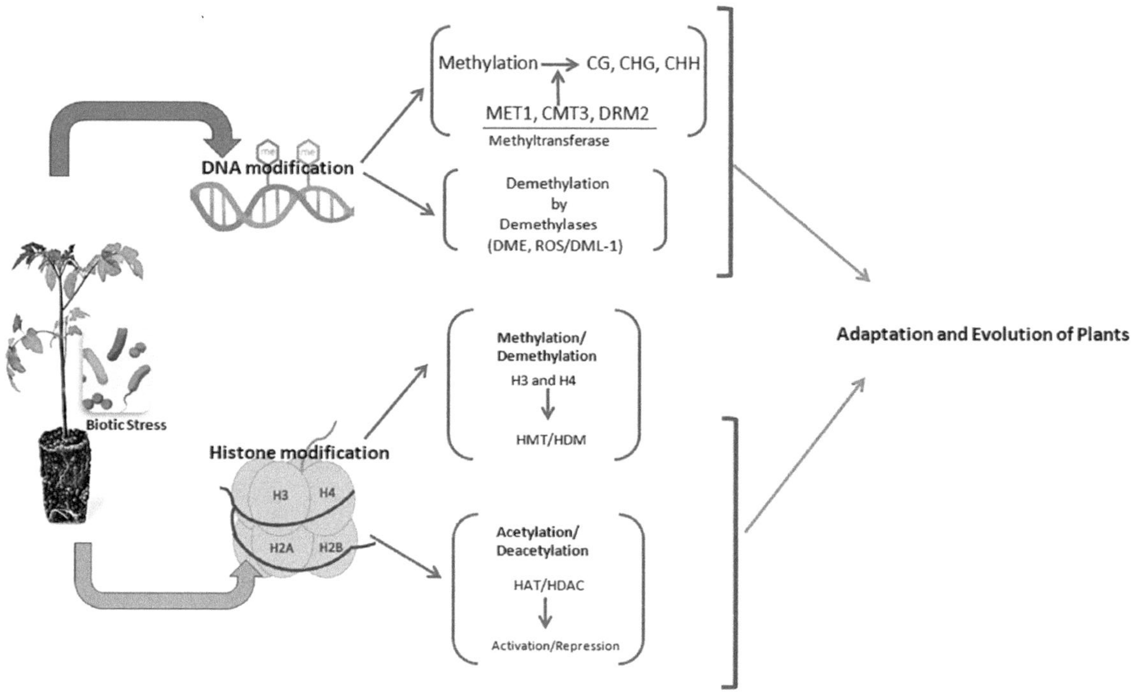

FIGURE 8.1 Epigenetic response during plant–pathogenic interaction.

Note: CG (or CpG); CHG; CHH (where H correspond to A, T or C); MET1: methyl transferase1; CMT3: chromo methylase3; DRM2: domains rearranged methyltransferase 2; DME: 5-methylcytosine DNA glycosylase/demethylase DEMETER (DME); ROS: reactive oxygen species; DML1: REPRESSOR OF SILENCING 1 (ROS1) also known as DEMETER-LIKE 1 (DML1); HMT: histone methyltransferases; HDM: histone demethylases; HAT: histone acetyltransferases; HDAC: histone deacetylase.

which methylation pathways link in the perspective of plant defense response. This was shown in a study on double mutants met1 nrpd2 that exhibited increased resistance to the bacterial pathogen *Pseudomonas syringae* (Pst) (Yu et al., 2013). In the same study, it was found that in the *Arabidopsis* plant, the main component RdDM pathway was down-regulated by bacterial flagellin-derived FLG 22, causing a lower level of CHH methylation. These findings reveal that DNA demethylation might influence rapid expression of defense-related genes during subsequent attack by the same or a different pathogen. Fungal pathogen *Hyaloperonospora arabidopsidis* (Hpa; downy mildew) infection in hypermethylated and hypomythylated *Arabidopsis* mutants was found to have an antagonistic effect. Hypermethylated ros1 mutants were more susceptible, while hypomethylated nrpe (defective RdDM) mutants were comparatively tolerant to Hpa (Lόpez Sánchez et al., 2016). Still, drawing straight relationships within epigenetic mutants and pleiotropic phenotypes is difficult.

Links between plant defense response and DNA methylation were observed in other host plants as well. Hypomethylated rice plants were observed showing more resistance to bacterial pathogen *Xanthomonas* after treatment with a demethylating agent 5-azadeoxycytidine (Dowen et al., 2012). In an experiment by Sha et al. (2005), rice seedlings and adult plants exposed to bacterial pathogen *Xanthomonas*, which causes blight disease, showed higher methylation levels in adult plants and varied methylation patterns. In bananas, after infection by *Fusarium oxysporum* f. sp. *cubense* tropical race 4 (Foc TR4), hypermethylation and hypomethylation happened at CCGG sites (CpG islands of promoters) in the whole genome (Luo et al., 2016). Here the DNA methylation pattern was notably changed after infection in bananas. Recently, it was found that infection of *Blumeria graminis* f. sp. *tritici* (Bgt) pathogen in wheat *Aegilops tauschii* decreases ARGONAUTE4a, 24-nt siRNAs levels and CHH methylation in pathogen responsive genes that are located near to TEs (Geng et al., 2019). There have been several studies published in recent years elucidating the role of DNA methylation in plant–nematode interactions (Hewezi, 2020; Hewezi et al., 2017; Rambani et al., 2020, 2015).

In plants, DNA methylation affects mitosis and meiosis processes by exerting effect on events like distribution of meiotic recombination hotspots, chromosome recombination events, allocation of the crossover events through chromosomes etc. (Choi et al., 2013; Mirouze et al., 2012; Yelina et al., 2012). DNA methylation might affect allele formation, which may lead to development of novel phenotypes and species, promoting evolution in plants, although it has been shown that the DNA methylation process is controlled by some evolutionary factors that considerably diverge across species, genes and methylated sequences (Takuno et al., 2016).

8.2.1.3 Post-Translational Histone Protein Modifications

The second key epigenetic process is post-translational modification of histone (PTMs) proteins (Berger et al., 2009). In plants, genomic DNA is packaged into a very organized nucleoprotein complex called a chromatin. Chromatins are made up of nucleosome units, and a single nucleosome unit is a compact structure of a total of 147 base pair DNA sequence wrapping around histone proteins. In a nucleosome, the core part has histone octamer, which contains two copies each of H2A, H2B, H3 and H4 histone protein. Modifications such as acetylation or methylation occur at the N terminal tail of histone, which is associated with gene activation or suppression events (He et al., 2011; Roudier et al., 2011). Several enzymes, like methyltransferases and demethylases, histone Acetyltransferases (HATs) and histone deacetylases (HDACs), mediate chemical modifications at N terminal tails (Fan et al., 2015). Acetylation of histone relaxes the chromatin structure so as to activate gene transcription, while methylation of histones can activate or suppress transcription activity based upon modified residues (Garner et al., 2016).

Both HATs and HDACs activity have been observed in the presence of biotic stresses. In *Arabidopsis*, mutation in ELON-GATOR PROTEIN 3 (which has acetyltransferase activity) initiated more sensitivity to *P. syringae pv. maculicola* and late defense gene transcription on pathogen infection (Defraia et al., 2013). Similarly, the ELON-GATOR PROTEIN 2 subunit is also needed for quick defense-related gene transcription after infection with pathogens (Defraia et al., 2013, 2010; C. Wang et al., 2015). In case of HDACs, several studies revealed that it acts as negative regulator of defense-related genes, and HDACs mutated plants have exhibited improved resistance to pathogen attacks (Choi et al., 2012; Ding & Wang, 2015; Espinas et al., 2016; C. Wang et al., 2010; Y. Wang et al., 2017). In rice, suppressing HDT 701 (histone H4 deacetylase) induced defense gene transcription and development of resistance to bacterial pathogens *Magnaporthe oryzae* and *Xanthomonas oryzae pv. oryzae* (Ding et al., 2012). In *Arabidopsis*, hda9hos15 (HAD 9 and its interacting protein HOS15 mutants) double mutant exhibited increased resistance to bacterial pathogen *Pseudomonas syringae* (Pst) (Yang et al., 2020).

In plants, histone methyltransferase and demethylase can play either positive or negative regulators of defense-related genes upon pathogen attacks. The rice demethylase OsJM704 acts as positive regulator of defense by suppressing the transcription of negative regulators of defense-related genes. In *Arabidopsis*, demethylases JUMONJI 27 (AtJMJ27), tJMJ14, and INCREASE IN BONSAI METHYLATION 1 are positive regulators and involved in expression of defense-related genes (Chan & Zimmerli, 2019; Dutta et al., 2017; Li et al., 2020). The methyltransferases SDG8 and SDG25 are also involved in developing resistance against biotic stresses, as sdg8 and sdg25 single and sdg8 sdg25 double mutants have been reported as showing more susceptibility to both fungal pathogen *Botrytis cinerea* and bacteria pathogen Pst (De-La-Peña et al., 2012; Lee et al., 2016). Recently, various studies have provided details about chromatin remodeling roles in plant defense mechanisms against pathogens (Berriri et al., 2016; Bordiya et al., 2016; Walley et al., 2008). In *Arabidopsis*, for example, the chromatin remodeling ATPase SPLAYED positively regulates resistance to necrotrophs. This ATPase also controls JA/ET response and plays a significant role against *Botrytis cinerea*, although it is not needed in developing resistance to Pst (Lee et al., 2016).

8.2.1.4 Small RNAs Mediated Modifications

Non-coding small RNAs play a key role in protecting plant systems from various pathogens by influencing expression of defense-related genes. Non-coding RNAs play significant role in defense against viruses as well as other pathogens. The movement of sRNAs from host to pathogen can trigger the silencing of genes. In *Arabidopsis*, sRNAs containing vesicles were gathered at the infection site after an attack of *Botrytis cinerea*. These vesicles fused with pathogen cells and delivered their matter into them, and the sRNAs suppress the expression of infection-related fungal genes (Schandry & Becker, 2020). However, pathogens also evolve and learn, using this strategy to suppress host plant immunity. In *Arabidopsis*, for instance, fungal pathogen oomycete Hpa transfers sRNAs into host cells and uses host ARGONAUTE1 protein to silence plant immune response (Dunker et al., 2020). In another study on *Arabidopsis*, it was observed that after infection with fungal pathogen *Phytophthora capsica*, locus-specific sRNAs were produced that suppress fungal infection-related genes, although the pathogen also reacts and interferes with siRNA biogenesis that affects plant defense response (Hou et al., 2019). Mobile mRNAs and miRNAs movement from one plant to another also affect plant immunity against pathogen attacks (Alakonya et al., 2012; Kim et al., 2014; LeBlanc et al., 2013). Although there are studies available about sRNA playing role across plants, its exact role in plant–pathogen interactions is still not clear.

8.3 CONCLUSION

Epigenetic interactions have emerged as major regulatory mechanisms upon exploring plants and their biotic stress-related responses. There are several studies available from the last two decades that reveal the role of epigenetic regulations to help plants in dealing with defense and symbiosis responses. DNA methylation, histone modifications and non-coding sRNA-mediated processes can act individually or together control the transcription activity of defense-related genes in plants. Although there is limited understanding about the molecular dynamics of interactions in these different epigenetic mechanisms, most studies have focused on the role of some epigenetic gene targets and epigenetic conformation of host cells, while genome-wide analysis of epigenetic modifications and epigenetic changes in pathogens can provide a better understanding of epigenetic regulations. Another limitation is the difficulty in comparing different studies because of experiment setup and pathogens are different. Studies seeking details about epigenetic modifications and their role in stress tolerance have revealed evolutionary developments of some species under specific conditions, to a certain level; however, there is still lack of sufficient information available depicting a clear picture of epigenetic modification's role in evolutionary adaptation events.

REFERENCES

Agius, F., Kapoor, A., & Zhu, J. K. (2006). Role of the *Arabidopsis* DNA glycosylase/lyase ROS1 in active DNA demethylation. *Proceedings of the National Academy of Sciences of the United States of America, 103*(31), 11796–11801. https://doi.org/10.1073/pnas.0603563103

Akhter, Z., Bi, Z., Ali, K., Sun, C., Fiaz, S., Haider, F. U., & Bai, J. (2021). In response to abiotic stress, DNA methylation confers epigenetic changes in plants. *Plants (Basel), 10*(6). https://doi.org/10.3390/plants10061096

Akimoto, K., Katakami, H., Kim, H. J., Ogawa, E., Sano, C. M., Wada, Y., & Sano, H. (2007). Epigenetic inheritance in rice plants. *Annals of Botany, 100*(2), 205–217. https://doi.org/10.1093/aob/mcm110

Alakonya, A., Kumar, R., Koenig, D., Kimura, S., Townsley, B., Runo, S., Garces, H. M., Kang, J., Yanez, A., David-Schwartz, R., Machuka, J., & Sinha, N. (2012). Interspecific RNA interference of shoot meristemless-like disrupts Cuscuta pentagona plant parasitism. *Plant Cell, 24*(7), 3153–3166. https://doi.org/10.1105/tpc.112.099994

Al-Lawati, A., Al-Bahry, S., Victor, R., Al-Lawati, A. H., & Yaish, M. W. (2016). Salt stress alters DNA methylation levels in alfalfa (Medicago spp). *Genetics and Molecular Research, 15*(1), 15018299. https://doi.org/10.4238/gmr.15018299

Baulcombe, D. C., & Dean, C. (2014). Epigenetic regulation in plant responses to the environment. *Cold Spring Harbor Perspectives in Biology, 6*(9), a019471. https://doi.org/10.1101/cshperspect.a019471

Bender, J. (2004). DNA methylation and epigenetics. *Annual Review of Plant Biology, 55*, 41–68. https://doi.org/10.1146/annurev.arplant.55.031903.141641

Berger, S. L., Kouzarides, T., Shiekhattar, R., & Shilatifard, A. (2009). An operational definition of epigenetics. *Genes & Development, 23*(7), 781–783. https://doi.org/10.1101/gad.1787609

Berriri, S., Gangappa, S. N., & Kumar, S. V. (2016). SWR1 chromatin-remodeling complex subunits and H2A.Z have non-overlapping functions in immunity and gene regulation in *Arabidopsis*. *Molecular Plant, 9*(7), 1051–1065. https://doi.org/10.1016/j.molp.2016.04.003

Bewick, A. J., Ji, L., Niederhuth, C. E., Willing, E. M., Hofmeister, B. T., Shi, X., Wang, L., Lu, Z., Rohr, N. A., Hartwig, B., Kiefer, C., Deal, R. B., Schmutz, J., Grimwood, J., Stroud, H., Jacobsen, S. E., Schneeberger, K., Zhang, X., & Schmitz, R. J. (2016). On the origin and evolutionary consequences of gene body DNA methylation. *Proceedings of the National Academy of Sciences of the United States of America, 113*(32), 9111–9116. https://doi.org/10.1073/pnas.1604666113

Bird, A. (2002). DNA methylation patterns and epigenetic memory. *Genes & Development, 16*(1), 6–21. https://doi.org/10.1101/gad.947102

Bochtler, M., Kolano, A., & Xu, G. L. (2017). DNA demethylation pathways: Additional players and regulators. *Bioessays, 39*(1), 1–13. https://doi.org/10.1002/bies.201600178

Bordiya, Y., Zheng, Y., Nam, J. C., Bonnard, A. C., Choi, H. W., Lee, B. K., Kim, J., Klessig, D. F., Fei, Z., & Kang, H. G. (2016). Pathogen infection and MORC proteins affect chromatin accessibility of transposable elements and expression of their proximal genes in *Arabidopsis*. *Molecular Plant-Microbe Interactions, 29*(9), 674–687. https://doi.org/10.1094/mpmi-01-16-0023-r

Bostick, M., Kim, J. K., Estève, P. O., Clark, A., Pradhan, S., & Jacobsen, S. E. (2007). UHRF1 plays a role in maintaining DNA methylation in mammalian cells. *Science, 317*(5845), 1760–1764. https://doi.org/10.1126/science.1147939

Chan, C., & Zimmerli, L. (2019). The histone demethylase IBM1 positively regulates *Arabidopsis* immunity by control of defense gene expression. *Frontiers in Plant Science, 10*, 1587. https://doi.org/10.3389/fpls.2019.01587

Chan, S. W., Henderson, I. R., & Jacobsen, S. E. (2005). Gardening the genome: DNA methylation in *Arabidopsis thaliana. Nature Reviews Genetics*, *6*(5), 351–360. https://doi.org/10.1038/nrg1601

Chen, L. J., Wei, S. Y., & Chiu, J. J. (2013). Mechanical regulation of epigenetics in vascular biology and pathobiology. *Journal of Cellular and Molecular Medicine*, *17*(4), 437–448. https://doi.org/10.1111/jcmm.12031

Choi, K., Zhao, X., Kelly, K. A., Venn, O., Higgins, J. D., Yelina, N. E., Hardcastle, T. J., Ziolkowski, P. A., Copenhaver, G. P., Franklin, F. C., McVean, G., & Henderson, I. R. (2013). *Arabidopsis* meiotic crossover hot spots overlap with H2A.Z nucleosomes at gene promoters. *Nature Genetics*, *45*(11), 1327–1336. https://doi.org/10.1038/ng.2766

Choi, S. M., Song, H. R., Han, S. K., Han, M., Kim, C. Y., Park, J., Lee, Y. H., Jeon, J. S., Noh, Y. S., & Noh, B. (2012). HDA19 is required for the repression of salicylic acid biosynthesis and salicylic acid-mediated defense responses in *Arabidopsis*. *Plant Journal*, *71*(1), 135–146. https://doi.org/10.1111/j.1365-313X.2012.04977.x

De-La-Peña, C., Rangel-Cano, A., & Alvarez-Venegas, R. (2012). Regulation of disease-responsive genes mediated by epigenetic factors: Interaction of *Arabidopsis*-Pseudomonas. *Molecular Plant Pathology*, *13*(4), 388–398. https://doi.org/10.1111/j.1364-3703.2011.00757.x

DeFraia, C. T., Wang, Y., Yao, J., & Mou, Z. (2013). Elongator subunit 3 positively regulates plant immunity through its histone acetyltransferase and radical S-adenosylmethionine domains. *BMC Plant Biology*, *13*, 102. https://doi.org/10.1186/1471-2229-13-102

DeFraia, C. T., Zhang, X., & Mou, Z. (2010). Elongator subunit 2 is an accelerator of immune responses in *Arabidopsis* thaliana. *Plant Journal*, *64*(3), 511–523. https://doi.org/10.1111/j.1365-313X.2010.04345.x

Ding, B., Bellizzi Mdel, R., Ning, Y., Meyers, B. C., & Wang, G. L. (2012). HDT701, a histone H4 deacetylase, negatively regulates plant innate immunity by modulating histone H4 acetylation of defense-related genes in rice. *Plant Cell*, *24*(9), 3783–3794. https://doi.org/10.1105/tpc.112.101972

Ding, B., & Wang, G. L. (2015). Chromatin versus pathogens: The function of epigenetics in plant immunity. *Frontiers in Plant Science*, *6*, 675. https://doi.org/10.3389/fpls.2015.00675

Douma, J. C., Vermeulen, P. J., Poelman, E. H., Dicke, M., & Anten, N. P. R. (2017). When does it pay off to prime for defense? A modeling analysis. *New Phytologist*, *216*(3), 782–797. https://doi.org/10.1111/nph.14771

Dowen, R. H., Pelizzola, M., Schmitz, R. J., Lister, R., Dowen, J. M., Nery, J. R., Dixon, J. E., & Ecker, J. R. (2012). Widespread dynamic DNA methylation in response to biotic stress. *Proceedings of the National Academy of Sciences of the United States of America*, *109*(32), E2183–E2191. https://doi.org/10.1073/pnas.1209329109

Dunker, F., Trutzenberg, A., Rothenpieler, J. S., Kuhn, S., Pröls, R., Schreiber, T., Tissier, A., Kemen, A., Kemen, E., Hückelhoven, R., & Weiberg, A. (2020). Oomycete small RNAs bind to the plant RNA-induced silencing complex for virulence. *Elife*, *9*. https://doi.org/10.7554/eLife.56096

Dutta, A., Choudhary, P., Caruana, J., & Raina, R. (2017). JMJ27, an *Arabidopsis* H3K9 histone demethylase, modulates defense against Pseudomonas syringae and flowering time. *Plant Journal*, *91*(6), 1015–1028. https://doi.org/10.1111/tpj.13623

Espinas, N. A., Saze, H., & Saijo, Y. (2016). Epigenetic control of defense signaling and priming in plants. *Frontiers in Plant Science*, *7*, 1201. https://doi.org/10.3389/fpls.2016.01201

Fan, J., Krautkramer, K. A., Feldman, J. L., & Denu, J. M. (2015). Metabolic regulation of histone post-translational modifications. *ACS Chemical Biology*, *10*(1), 95–108. https://doi.org/10.1021/cb500846u

Ficz, G., Hore, T. A., Santos, F., Lee, H. J., Dean, W., Arand, J., Krueger, F., Oxley, D., Paul, Y. L., Walter, J., Cook, S. J., Andrews, S., Branco, M. R., & Reik, W. (2013). FGF signaling inhibition in ESCs drives rapid genome-wide demethylation to the epigenetic ground state of pluripotency. *Cell Stem Cell*, *13*(3), 351–359. https://doi.org/10.1016/j.stem.2013.06.004

Garner, C. M., Kim, S. H., Spears, B. J., & Gassmann, W. (2016). Express yourself: Transcriptional regulation of plant innate immunity. *Seminars in Cell and Developmental Biology*, *56*, 150–162. https://doi.org/10.1016/j.semcdb.2016.05.002

Geng, S., Kong, X., Song, G., Jia, M., Guan, J., Wang, F., Qin, Z., Wu, L., Lan, X., Li, A., & Mao, L. (2019). DNA methylation dynamics during the interaction of wheat progenitor Aegilops tauschii with the obligate biotrophic fungus Blumeria graminis f. sp. tritici. *New Phytologist*, *221*(2), 1023–1035. https://doi.org/10.1111/nph.15432

Gouil, Q., & Baulcombe, D. C. (2016). DNA methylation signatures of the plant chromomethyltransferases. *PLoS Genetics*, *12*(12), e1006526. https://doi.org/10.1371/journal.pgen.1006526

Grant-Downton, R. T., & Dickinson, H. G. (2005). Epigenetics and its implications for plant biology. 1: The epigenetic network in plants. *Annals of Botany*, *96*(7), 1143–1164. https://doi.org/10.1093/aob/mci273

He, G., Elling, A. A., & Deng, X. W. (2011). The epigenome and plant development. *Annual Review of Plant Biology*, *62*, 411–435. https://doi.org/10.1146/annurev-arplant-042110-103806

Henderson, I. R., & Jacobsen, S. E. (2007). Epigenetic inheritance in plants. *Nature*, *447*(7143), 418–424. https://doi.org/10.1038/nature05917

Hewezi, T. (2020). Epigenetic mechanisms in nematode-plant interactions. *Annual Review of Phytopathology*, *58*, 119–138. https://doi.org/10.1146/annurev-phyto-010820-012805

Hewezi, T., Lane, T., Piya, S., Rambani, A., Rice, J. H., & Staton, M. (2017). Cyst nematode parasitism induces dynamic changes in the root epigenome. *Plant Physiology*, *174*(1), 405–420. https://doi.org/10.1104/pp.16.01948

Hewezi, T., Pantalone, V., Bennett, M., Neal Stewart, C., Jr., & Burch-Smith, T. M. (2018). Phytopathogen-induced changes to plant methylomes. *Plant Cell Reports*, *37*(1), 17–23. https://doi.org/10.1007/s00299-017-2188-y

Hirayama, T., & Shinozaki, K. (2010). Research on plant abiotic stress responses in the post-genome era: Past, present and future. *Plant J*, *61*(6), 1041–1052. https://doi.org/10.1111/j.1365-313X.2010.04124.x

Hirsch, C. N., Hirsch, C. D., Felcher, K., Coombs, J., Zarka, D., Van Deynze, A., De Jong, W., Veilleux, R. E., Jansky, S., Bethke, P., Douches, D. S., & Buell, C. R. (2013). Retrospective view of North American potato (Solanum tuberosum L.) breeding in the 20th and 21st centuries. *G3 (Bethesda)*, *3*(6), 1003–1013. https://doi.org/10.1534/g3.113.005595

Hou, Y., Zhai, Y., Feng, L., Karimi, H. Z., Rutter, B. D., Zeng, L., Choi, D. S., Zhang, B., Gu, W., Chen, X., Ye, W., Innes, R. W., Zhai, J., & Ma, W. (2019). A phytophthora effector suppresses trans-kingdom RNAi to promote disease susceptibility. *Cell Host Microbe*, *25*(1), 153–165.e155. https://doi.org/10.1016/j.chom.2018.11.007

Kim, G., LeBlanc, M. L., Wafula, E. K., dePamphilis, C. W., & Westwood, J. H. (2014). Plant science: Genomic-scale exchange of mRNA between a parasitic plant and its hosts. *Science*, *345*(6198), 808–811. https://doi.org/10.1126/science.1253122

Klose, R. J., & Bird, A. P. (2006). Genomic DNA methylation: The mark and its mediators. *Trends in Biochemical Sciences*, *31*(2), 89–97. https://doi.org/10.1016/j.tibs.2005.12.008

Kumar, S. (2018). Epigenomics of plant responses to environmental stress. *Epigenomes*, *2*(1), 6.

Kumar, S., & Singh, A. (2016). Epigenetic regulation of abiotic stress tolerance in plants. *Advances in Plants & Agriculture Research*, *5*, 00179.

Laland, K., Uller, T., Feldman, M., Sterelny, K., Müller, G. B., Moczek, A., Jablonka, E., Odling-Smee, J., Wray, G. A., Hoekstra, H. E., Futuyma, D. J., Lenski, R. E., Mackay, T. F., Schluter, D., & Strassmann, J. E. (2014). Does evolutionary theory need a rethink? *Nature*, *514*(7521), 161–164. https://doi.org/10.1038/514161a

LeBlanc, M., Kim, G., Patel, B., Stromberg, V., & Westwood, J. (2013). Quantification of tomato and *Arabidopsis* mobile RNAs trafficking into the parasitic plant Cuscuta pentagona. *New Phytologist*, *200*(4), 1225–1233. https://doi.org/10.1111/nph.12439

Lee, S., Fu, F., Xu, S., Lee, S. Y., Yun, D. J., & Mengiste, T. (2016). Global regulation of plant immunity by histone lysine methyl transferases. *Plant Cell*, *28*(7), 1640–1661. https://doi.org/10.1105/tpc.16.00012

Lennartsson, A. (2017). Dynamics and function of DNA methylation during development. In *Chromatin regulation and dynamics* (pp. 65–94). Elsevier.

Li, D., Liu, R., Singh, D., Yuan, X., Kachroo, P., & Raina, R. (2020). JMJ14 encoded H3K4 demethylase modulates immune responses by regulating defence gene expression and pipecolic acid levels. *New Phytologist*, *225*(5), 2108–2121. https://doi.org/10.1111/nph.16270

López Sánchez, A., Stassen, J. H., Furci, L., Smith, L. M., & Ton, J. (2016). The role of DNA (de)methylation in immune responsiveness of *Arabidopsis*. *Plant Journal*, *88*(3), 361–374. https://doi.org/10.1111/tpj.13252

Luo, J.-Y., Pan, X.-L., Peng, T.-C., Chen, Y.-Y., Hui, Z., Lei, M., Yun, P., Rui, H., & Hua, T. (2016). DNA methylation patterns of banana leaves in response to Fusarium oxysporum f. sp. cubense tropical race 4. *Journal of Integrative Agriculture*, *15*(12), 2736–2744.

Mirouze, M., Lieberman-Lazarovich, M., Aversano, R., Bucher, E., Nicolet, J., Reinders, J., & Paszkowski, J. (2012). Loss of DNA methylation affects the recombination landscape in *Arabidopsis*. *Proceedings of the National Academy of Sciences of the United States of America*, *109*(15), 5880–5885. https://doi.org/10.1073/pnas.1120841109

Niederhuth, C. E., & Schmitz, R. J. (2017). Putting DNA methylation in context: From genomes to gene expression in plants. *Biochimica et Biophysica Acta: Gene Regulatory Mechanisms*, *1860*(1), 149–156. https://doi.org/10.1016/j.bbagrm.2016.08.009

Niehrs, C. (2009). Active DNA demethylation and DNA repair. *Differentiation*, *77*(1), 1–11. https://doi.org/10.1016/j.diff.2008.09.004

Paran, I., & van der Knaap, E. (2007). Genetic and molecular regulation of fruit and plant domestication traits in tomato and pepper. *Journal of Experimental Botany*, *58*(14), 3841–3852. https://doi.org/10.1093/jxb/erm257

Piccolo, F. M., & Fisher, A. G. (2014). Getting rid of DNA methylation. *Trends in Cell Biology*, *24*(2), 136–143. https://doi.org/10.1016/j.tcb.2013.09.001

Pigliucci, M. (2007). Do we need an extended evolutionary synthesis? *Evolution*, *61*(12), 2743–2749. https://doi.org/10.1111/j.1558-5646.2007.00246.x

Pikaard, C. S., Haag, J. R., Ream, T., & Wierzbicki, A. T. (2008). Roles of RNA polymerase IV in gene silencing. *Trends in Plant Science*, *13*(7), 390–397. https://doi.org/10.1016/j.tplants.2008.04.008

Rambani, A., Hu, Y., Piya, S., Long, M., Rice, J. H., Pantalone, V., & Hewezi, T. (2020). Identification of differentially methylated miRNA genes during compatible and incompatible interactions between soybean and soybean cyst nematode. *Molecular Plant-Microbe Interactions*, *33*(11), 1340–1352. https://doi.org/10.1094/mpmi-07-20-0196-r

Rambani, A., Rice, J. H., Liu, J., Lane, T., Ranjan, P., Mazarei, M., Pantalone, V., Stewart, C. N., Staton, M., & Hewezi, T. (2015). The methylome of soybean roots during the compatible interaction with the soybean cyst nematode. *Plant Physiology*, *168*(4), 1364–1377. https://doi.org/10.1104/pp.15.00826

Ramirez-Prado, J. S., Piquerez, S. J. M., Bendahmane, A., Hirt, H., Raynaud, C., & Benhamed, M. (2018). Modify the histone to win the battle: Chromatin dynamics in plant-pathogen interactions. *Frontiers in Plant Science*, *9*, 355. https://doi.org/10.3389/fpls.2018.00355

Ramos-Cruz, D., Troyee, A. N., & Becker, C. (2021). Epigenetics in plant organismic interactions. *Current Opinion in Plant Biology*, *61*, 102060. https://doi.org/10.1016/j.pbi.2021.102060

Richards, C. L., Bossdorf, O., & Verhoeven, K. J. (2010). Understanding natural epigenetic variation. *New Phytologist*, *187*(3), 562–564. https://doi.org/10.1111/j.1469-8137.2010.03369.x

Richards, E. J. (2006). Inherited epigenetic variation – revisiting soft inheritance. *Nature Reviews Genetics*, *7*(5), 395–401. https://doi.org/10.1038/nrg1834

Richards, E. J. (2011). Natural epigenetic variation in plant species: A view from the field. *Current Opinion in Plant Biology*, *14*(2), 204–209. https://doi.org/10.1016/j.pbi.2011.03.009

Roudier, F., Ahmed, I., Bérard, C., Sarazin, A., Mary-Huard, T., Cortijo, S., Bouyer, D., Caillieux, E., Duvernois-Berthet, E., Al-Shikhley, L., Giraut, L., Després, B., Drevensek, S., Barneche, F., Dèrozier, S., Brunaud, V., Aubourg, S., Schnittger, A., Bowler, C., . . . Colot, V. (2011). Integrative epigenomic mapping defines four main chromatin states in *Arabidopsis*. *Embo Journal*, *30*(10), 1928–1938. https://doi.org/10.1038/emboj.2011.103

Saze, H. (2008). Epigenetic memory transmission through mitosis and meiosis in plants. *Seminars in Cell and Developmental Biology*, *19*(6), 527–536. https://doi.org/10.1016/j.semcdb.2008.07.017

Schandry, N., & Becker, C. (2020). Allelopathic plants: Models for studying plant-interkingdom interactions. *Trends in Plant Science*, *25*(2), 176–185. https://doi.org/10.1016/j.tplants.2019.11.004

Schmitz, R. J., & Ecker, J. R. (2012). Epigenetic and epigenomic variation in *Arabidopsis* thaliana. *Trends in Plant Science*, *17*(3), 149–154. https://doi.org/10.1016/j.tplants.2012.01.001

Sha, A. H., Lin, X. H., Huang, J. B., & Zhang, D. P. (2005). Analysis of DNA methylation related to rice adult plant resistance to bacterial blight based on methylation-sensitive AFLP (MSAP) analysis. *Molecular Genetics and Genomics*, *273*(6), 484–490. https://doi.org/10.1007/s00438-005-1148-3

Simon, S. A., & Meyers, B. C. (2011). Small RNA-mediated epigenetic modifications in plants. *Current Opinion in Plant Biology*, *14*(2), 148–155. https://doi.org/10.1016/j.pbi.2010.11.007

Slotkin, R. K. (2016). Plant epigenetics: From genotype to phenotype and back again. *Genome Biology*, *17*, 57. https://doi.org/10.1186/s13059-016-0920-5

Stassen, J. H. M., López, A., Jain, R., Pascual-Pardo, D., Luna, E., Smith, L. M., & Ton, J. (2018). The relationship between transgenerational acquired resistance and global DNA methylation in *Arabidopsis*. *Scientific Reports*, *8*(1), 14761. https://doi.org/10.1038/s41598-018-32448-5

Stroud, H., Do, T., Du, J., Zhong, X., Feng, S., Johnson, L., Patel, D. J., & Jacobsen, S. E. (2014). Non-CG methylation patterns shape the epigenetic landscape in *Arabidopsis*. *Nature Structural & Molecular Biology*, *21*(1), 64–72. https://doi.org/10.1038/nsmb.2735

Stroud, H., Greenberg, M. V., Feng, S., Bernatavichute, Y. V., & Jacobsen, S. E. (2013). Comprehensive analysis of silencing mutants reveals complex regulation of the *Arabidopsis* methylome. *Cell*, *152*(1–2), 352–364. https://doi.org/10.1016/j.cell.2012.10.054

Takuno, S., Ran, J. H., & Gaut, B. S. (2016). Evolutionary patterns of genic DNA methylation vary across land plants. *Nature Plants*, *2*, 15222. https://doi.org/10.1038/nplants.2015.222

Thiebaut, F., Hemerly, A. S., & Ferreira, P. C. G. (2019). A role for epigenetic regulation in the adaptation and stress responses of non-model plants. *Frontiers in Plant Science*, *10*, 246. https://doi.org/10.3389/fpls.2019.00246

Vaissière, T., Sawan, C., & Herceg, Z. (2008). Epigenetic interplay between histone modifications and DNA methylation in gene silencing. *Mutation Research*, *659*(1–2), 40–48. https://doi.org/10.1016/j.mrrev.2008.02.004

Walley, J. W., Rowe, H. C., Xiao, Y., Chehab, E. W., Kliebenstein, D. J., Wagner, D., & Dehesh, K. (2008). The chromatin remodeler SPLAYED regulates specific stress signaling pathways. *PLoS Pathogens*, *4*(12), e1000237. https://doi.org/10.1371/journal.ppat.1000237

Wang, C., Ding, Y., Yao, J., Zhang, Y., Sun, Y., Colee, J., & Mou, Z. (2015). *Arabidopsis* Elongator subunit 2 positively contributes to resistance to the necrotrophic fungal pathogens Botrytis cinerea and Alternaria brassicicola. *Plant Journal*, *83*(6), 1019–1033. https://doi.org/10.1111/tpj.12946

Wang, C., Gao, F., Wu, J., Dai, J., Wei, C., & Li, Y. (2010). *Arabidopsis* putative deacetylase AtSRT2 regulates basal defense by suppressing PAD4, EDS5 and SID2 expression. *Plant & Cell Physiology*, *51*(8), 1291–1299. https://doi.org/10.1093/pcp/pcq087

Wang, Y., Hu, Q., Wu, Z., Wang, H., Han, S., Jin, Y., Zhou, J., Zhang, Z., Jiang, J., & Shen, Y. (2017). Histone deacetylase 6 represses pathogen defence responses in *Arabidopsis* thaliana. *Plant, Cell & Environment*, *40*(12), 2972–2986.

Wilschut, R. A., Oplaat, C., Snoek, L. B., Kirschner, J., & Verhoeven, K. J. (2016). Natural epigenetic variation contributes to heritable flowering divergence in a widespread asexual dandelion lineage. *Molecular Ecology*, *25*(8), 1759–1768. https://doi.org/10.1111/mec.13502

Wu, C., & Morris, J. R. (2001). Genes, genetics, and epigenetics: A correspondence. *Science*, *293*(5532), 1103–1105. https://doi.org/10.1126/science.293.5532.1103

Xia, H., Huang, W., Xiong, J., Tao, T., Zheng, X., Wei, H., Yue, Y., Chen, L., & Luo, L. (2016). Adaptive epigenetic differentiation between upland and lowland rice ecotypes revealed by methylation-sensitive amplified polymorphism. *PLoS One*, *11*(7), e0157810. https://doi.org/10.1371/journal.pone.0157810

Yaish, M. W. (2013). DNA methylation-associated epigenetic changes in stress tolerance of plants. In *Molecular stress physiology of plants* (pp. 427–440). Springer.

Yaish, M. W., Patankar, H. V., Assaha, D. V. M., Zheng, Y., Al-Yahyai, R., & Sunkar, R. (2017). Genome-wide expression profiling in leaves and roots of date palm (Phoenix dactylifera L.) exposed to salinity. *BMC Genomics*, *18*(1), 246. https://doi.org/10.1186/s12864-017-3633-6

Yang, X., Sanchez, R., Kundariya, H., Maher, T., Dopp, I., Schwegel, R., Virdi, K., Axtell, M. J., & Mackenzie, S. A. (2020). Segregation of an MSH1 RNAi transgene produces heritable non-genetic memory in association with methylome reprogramming. *Nature Communications*, *11*(1), 2214. https://doi.org/10.1038/s41467-020-16036-8

Yelina, N. E., Choi, K., Chelysheva, L., Macaulay, M., de Snoo, B., Wijnker, E., Miller, N., Drouaud, J., Grelon, M., Copenhaver, G. P., Mezard, C., Kelly, K. A., & Henderson, I. R. (2012). Epigenetic remodeling of meiotic crossover frequency in *Arabidopsis* thaliana DNA methyltransferase mutants. *PLoS Genetics*, *8*(8), e1002844. https://doi.org/10.1371/journal.pgen.1002844

Yu, A., Lepère, G., Jay, F., Wang, J., Bapaume, L., Wang, Y., Abraham, A. L., Penterman, J., Fischer, R. L., Voinnet, O., & Navarro, L. (2013). Dynamics and biological relevance of DNA demethylation in *Arabidopsis* antibacterial defense. *Proceedings of the National Academy of Sciences of the United States of America*, *110*(6), 2389–2394. https://doi.org/10.1073/pnas.1211757110

Zhang, H., Lang, Z., & Zhu, J. K. (2018). Dynamics and function of DNA methylation in plants. *Nature Reviews Molecular Cell Biology*, *19*(8), 489–506. https://doi.org/10.1038/s41580-018-0016-z

Zhao, Y., & Chen, X. (2014). Noncoding RNAs and DNA methylation in plants. *National Science Review*, *1*(2), 219–229. https://doi.org/10.1093/nsr/nwu003

Zhu, Q. H., Shan, W. X., Ayliffe, M. A., & Wang, M. B. (2016). Epigenetic mechanisms: An emerging player in plant-microbe interactions. *Molecular Plant-Microbe Interactions*, *29*(3), 187–196. https://doi.org/10.1094/mpmi-08-15-0194-fi

9 Gene Regulation for Drought, Cold, Heavy Metal and Environmental Responses

Aneeta Tiwari, Elhan Khan, Sonam Dwivedi, Haram Sarfraz, and Iffat Zareen Ahmad

9.1 INTRODUCTION

By modification of the CBF pathway, cold tolerance can be enhanced in different plant species. Post-transcriptional control of pre-mRNA processing and nucleus export is critical for cold adaptation. For post-transcriptional gene silencing, microRNAs (miRNAs) and small interfering RNAs (siRNAs) are essential (Chinnusamy et al., 2007). Cold stress, a fundamental environmental stress, has an effect on crop yield, quality, and post-harvest plant life. During the cold acclimation process, many temperate plants gain freezing and chilling resistance after being exposed to sublethal cold stress. However, many agronomical crops are unable to do so. During cold acclimation, the expression of transcriptional regulators and effector genes, also referred to as cold-regulated (COR) genes, is exceptionally controlled. The expedition for transcriptional, post-transcriptional, and post-translational expression of COR gene regulators in response to cold-induced stress has been reported (Yamaguchi-Shinozaki & Shinozaki, 2006).

Freezing tolerance-based genetic screening has identified a number of intriguing *Arabidopsis* mutants with improved or decreased freezing tolerance. Cold stress can produce changes in membrane fluidity and protein structure in plant cells. Rigidification of the cell membrane induced via stress at microdomains may cause actin cytoskeletal reorganization followed by Ca2+ channel activation and a rise in cytosolic Ca2+ level, leading to COR gene expression during cold acclimation. Los5 is unable to establish freezing tolerance and has a large drop in COR gene expression driven by cold and salt/drought (RD29A, COR15, COR47, RD22, and pyrroline-5-carboxylate synthase). As a result, ABA is required for plant cold tolerance (Xiong et al., 2001). Cold stress signals may be sent by ABA through second messengers like H2O2 and Ca2+ (Chinnusamy et al., 2006).

DOI: 10.1201/9781003153481-9

Drought is a harmful environmental factor that affects agriculture by preventing flora from attaining their full gene expression. In many parts of the world, salt stress affects plant agriculture, particularly on irrigated land. Drought affects agriculture and the economy more than salt stress. Drought stress signaling ought to be treated separately in this aspect. Because the reactions of plants to salt and drought are inextricably linked, and their pathways are also interconnected, there are more studies on water stress signaling. If we talk practically, salt stress can be applied more easily in laboratory circumstances (Zhu, 2002).The intricacy of transcriptional regulation is demonstrated by the large number of regulatory genes implicated in response to varied environmental stimuli and their overlapping expression spectra (Chen & Zhu, 2004).

Heavy metal stress has an impact on plant development. Heavy metals are made up of both non-essential and essential metals. When flora is exposed to heavy metals, it uses complex gene regulation mechanisms. MicroRNAs (non-coding short RNAs with 21 nucleotides) are a kind of non-coding RNA and are regulators of gene expression in heavy metals. MicroRNAs that are metal-regulated, as well as their target genes, have an impact on heavy metal absorption and transfer, metal chelation, protein folding, and reactive oxygen species scavenging, all of which are intertwined in a complicated regulatory network (Ding et al., 2013). The components of plant cell responses govern the unfavorable state at the post-transcriptional level, pointing to RNA molecules as one of the key players at this level of gene regulation. Plants cope with unfavorable situations such as biochemical, physiological, and molecular processes by employing a range of mechanisms at various levels (Covarrubias & Reyes, 2010).

Drought, cold, heavy metals, and environmental stress all have molecular reactions, and gene regulation is one of them. Understanding how trees respond physiologically and genetically to drought and other environmental conditions is essential for enhancing their growth, survival, and drought tolerance. When stress activates certain proteins involved in signal transduction pathways that are produced by specific genes, it leads to an increase in stress tolerance. On the other hand, the genetic underpinnings of stress resistance systems are complex and poorly understood. To manage the response to drought and salt stress, it is necessary to identify the relevant genes and define their regulation (Yoon et al., 2014).

A plant's production is significantly harmed by the effects of salt on respiration, nutritional absorption, hormonal imbalance, and other factors. Salt stress produces reactive oxygen species, which damage macromolecules such as nucleic acids, proteins and lipids, as well as reduce agricultural output. Expressed genes that are involved in the SOS signaling network, proteins, metabolites, transcription factors (MYB, HLH and many more), and metabolites (phytohormones, lipids, osmolytes) are all part of the plant salt-tolerant network and are utilized to modify plants with enhanced tolerance against salt and also to increase agricultural productivity. This method is quite beneficial in fulfilling the ever-increasing food demand (Parihar et al., 2015). Stress-responsive promoters in the signal transduction network contain a variety of transcription factors. These promoters serve as both molecular switches for gene expression and as stress signaling terminals in signaling events (Yamaguchi-Shinozaki & Shinozaki, 2006).

9.2 EFFECTS OF ENVIRONMENTAL STRESS ON PLANTS

The reinforcement of plants needs an ideal range of temperature. A specific set of temperature conditions exists, which may differ from plant to plant. Cold stress causes inadequate germination, reduced seedlings, browning of leaves, dying, and decreased seedling growth. Cold stress during the growing season induces plants to delay budding and pollination, which is the primary cause of crop output decreases. Plasma membrane damage is seen due to cold stress. Cold stress also induces dehydration and ice formation that damage the plasma membrane (Yadav, 2010). Abiotic stress factors are the biggest yield-reducing factors in the world, and they prevent the growth of plants and their germination (Table 9.1). Drought is a crop yield - limiting condition that hinders crop productivity. To develop this feature, which eventually manifests as "drought tolerance," breeding involves fundamental alterations in the collection of critical qualities. Drought is a serious environmental stressor in most areas around the world, particularly in hot and dry conditions with low crop yields. Plant growth is stressed by dryness, and its effects on the yield of a crop are lowered by 50–30% due to low humidity. This will lead to a lot of evaporation because the temperature is high. Drought-induced high temperatures promote respiration (Fathi & Tari, 2016). Heavy metals, such as copper, zinc, cadmium, and lead, produce heavy metal stress. Under stress conditions, ethylene promotes growth inhibition through increasing the rigidity of cell walls by enhancing the lignification process in stressful situations. Copper's action on dicotyledon plant roots and leaves may be inhibited by ethylene (Maksymiec, 2007). Peroxidase is an enzyme that can oxidize a variety of substrates in the presence of hydrogen peroxide. Several physiological and biochemical processes rely on these enzymes, including cell growth and expansion, differentiation, and abiotic and biotic stress responses. Iron toxicity in rice is caused by high ferrous iron in flooded soil, which stimulates peroxidase activity in rice leaves. Zn2+, Cd2+, Ni2+, and Cu2+ have been found to modify the isozyme patterns of peroxidase (Fang & Kao, 2000).

9.3 GENE REGULATION IN PLANTS IN DROUGHT STRESS

9.3.1 Effects of Flavonoids on Wheat During Drought

Flavonoids are physiologically active phytochemicals and play a vital role in development, breeding, maturation, and strain resistance. The actions of important enzymes involved in the flavonoid synthesis in winter wheat are poorly characterized (Ma et al., 2014). Flavonoids have a variety of functions, including serving as developmental regulators and responding to a variety of environmental stresses. It also aids in the reduction of short-wavelength UV-B radiation from the sun in the

280–315 nm region (Agati & Tattini, 2010). When drought stress occurs, flavonoid biosynthesis pathways play a modest role; flavonoids accumulate in wheat leaves, and expression of flavonoid biosynthetic pathways increases. The synthesis of flavonoids is an important process that overcomes the condition of drought. Chalcone synthase is the precursor of flavonoid pathways, and chalconeisomerase converted chalcone to flavanone, which switched on many other flavonoid molecules. The CHS gene had the highest expression of genes during periods of intense drought tolerance. In *Arabidopsis*, the interaction between CHS and CHI has been identified. Drought stress increases the activation of the TaF3H genes in wheat leaves, increasing drought resistance. Flavonoids have a protective role and efficiently resist stress-induced oxidative damage. Flavonoids, once formed, can act as antioxidants by preventing the formation of reactive oxygen species (Ma et al., 2014).

9.3.1.1 Action of Abscisic Acid Under Drought Stress

Abscisic acid (ABA) is a phytohormone that causes plants to become drought resistant. Exogenous ABA substantially activates the cellular metabolism of tea leaves when they are subjected to drought stress, including metabolic activities, amino acid metabolic activity, lipogenesis, and flavonoid production. Exogenous ABA enhanced the content of flavones, anthocyanins, flavonol, and isoflavone in tea leaves while the amount of lipid and fatty acids become reduced under stress. External ABA could promote in the recovery of tea leaves from drought stress by activating genes and altering metabolite content (Gai et al., 2020). The plant's reaction is regulated by ABA, which causes closure of the stomata and decreases moisture loss via limiting osmosis. When exposed to ABA, guard cells shrink in bulk and seal across the opening in the trachea. This can be accomplished by altering the flow of ions within the guard cell. Enhance flow velocity and ABA stimulates growth in the root, resulting in scarcity of water. The accumulation of isotonic effective compounds rate as a function of ABA, which safeguards cells from destruction (Daszkowska-Golec, 2016).

9.3.1.2 Downregulation of miRNA-Mediated Gene Regulation under Drought Stress

Introns typically disrupt microRNA (miRNA) binding sites (MBSs), necessitating a correction to produce the required MBSs in target transcripts, with splicing being used. As a result of alternative splicing, the miR398 binding site is lost in AhCSD1–2. During drought conditions, miRNA-mediated dysregulation is overcome. Peanut, soybean, and Arabidopsis contain the alternative isoform. Arabidopsis transgenic plants expressing AhCSD1 were likewise more resistant to osmotic stress. Increased in the level of miR398, AhCSD1 is enhanced to provide for a range of plant adaptations to combat ecological problems (Park & Grabau, 2017).

9.3.1.3 Changes in the Expression of Sugarcane Gene under Drought

Sugarcane is a valuable crop that generates around 65% of the world's sugar. Sugarcane is a polyploidy grass with a complicated genetic makeup. Due to the morphological impacts of drought, sugarcane productivity is reduced. There is a sugarcane genotype, RB72910, that is drought tolerant (Andrade et al., 2015). To improve abiotic stress tolerance, traditional breeding programs are being used. However, they are constrained by the trait's multigenic nature. Because of access to knock-out variants and its adaptability to gene transfer, *Arabidopsis thaliana* is essential in determining the activities of specific stress-associated genes (Bartels & Sunkar, 2005).

9.3.1.4 MicroRNAs in the Drought Stress

Drought reduces crop production. Understanding the basic mechanisms involved in this process is critical, as is selecting plants with high tolerance efficacy. Cowpea is animal feed crop grown in Asia, Africa, the United States, and Central America. Due to its drought tolerance, cowpea has been widely embraced in these areas. Cowpea contains drought-tolerant traits that are linked to the seedlings. MicroRNAs (miRNAs) have a length of 20–24 nucleotides (nt). RNA folds into stem-loop structures to form the single-stranded RNA molecules. Dehydration increased the expression of three miRNA 393, 397b, and 402, while decreasing the expression of miR389a. Drought stress tolerance in plants is mediated by these miRNAs. Drought stress was discovered to down-regulate miR169 relying on abscisic acid. MiRNA 169 gets suppressed, resulting in an increase in the expression of the NFYA5 gene (Ding et al., 2013).

9.3.1.5 The Gene Regulation Network Is Behind Drought Tolerance in Black Poplar (*Populusnigra* L.)

Among black poplar clones, there are drought-adaptive techniques. The sensitive genotype was designated as N.03.368, and it demonstrated drought-escape adaptation features. Under drought stress, it reduces leaf area to reduce transpiration, and it grows quickly when there is plenty of water. Drought resistance and tolerance adaptation are found in black poplar tree genotypes. Genes involved in planned leaf senescence are regulated. However, this genotype had upregulated the production of riboflavin GTP cyclohydrolase II and two transcription factors, WRKY and ERFs, that were linked to drought sensing and regulation (Yıldırım & Kaya, 2017).

9.3.1.6 Gene Regulation in Cold-Stressed Plants

Plants from temperate locations differ greatly in their aptitude for thriving in frigid temperatures. The ability of such plants to tolerate cold is limited. Progressively, environmental signals induce many sensory variations that indicate the upcoming winter and induce frost-tolerant properties. The phenomenon, called cold acclimation, is the pivotal element of the environment that is responsible for triggering frost tolerance in plants. Non-acclimated rye is damaged at approximately –5°C freezing temperature, but if exposed to lower non-freezing temperatures for an extended period of time, it can survive at temperatures as low as –30°C (Lee et al., 2002).

9.3.2 Plant Gene Expression Modulation in Response to Cold Stress

Plants under stress due to cold are reported to regulate their transcriptome by regulating expression of involved genes at transcription, post-transcription, and post-translation stages. Most of the plant's diversity, the cascade of ICE1–CBF transcriptional, provides a vital function toward cold acclimation, a key role in cold tolerance. Additionally, ICE1 seems to adversely influence MYB15, which is an upstream negative regulator of CBFs, in addition to directly inducing CBF expression. Furthermore, ICE1 expression is controlled post-translationally via proteolysis and ubiquitination induced by HOS1 and also SIZ1-mediated sumoylation, resulting in regulating the activity and level of ICE1. There are several CBF-independent regulons that are critical for cold acclimation. Chinnusamy and colleagues describe post-transcriptional regulatory mechanisms such as mRNA degradation, mRNA export, and pre-mRNA splicing directed by small RNAs as significant against frost tolerance (Chinnusamy et al., 2007). One of the transcriptome profiling studies in *Arabidopsis* under cold acclimation revealed the activation of multiple regulatory and CBF response pathways. Under stress, it induces the rapid expression of the DREB1 protein, which is a transcription activator of CBF. Furthermore, it induces the CBF-regulated CBF regulons, which are responsible for triggering tolerance against freezing temperatures (Figures 9.1 and 9.2). The seven-day study identified a total of 396 cold-responsive genes, and 218 showed increased transcript levels, while the remaining 88 genes showed threefold or more reduced transcripts. The analysis shows the significant downregulation of genes during cold acclimation (Fowler & Thomashow, 2002).

Similarly, another study screened an ice mutant of *Arabidopsis thaliana* that impedes CBF3 reporter gene transcription (promoter-luciferase) induced by cold. The mutation induced by cold prevents CBF3 expression and downregulates the expression of many CBF downstream genes, which is reported to be responsible for decreased tolerance toward cold temperatures and plant chilling. A transcriptional activator, MYC-like bHLH, is encoded by ICE1. The MYC-recognition regions in the CBF3 promoter are targeted by ICE1. When plants are exposed to cold temperatures, ICE1 is produced constitutively, and overexpression of ICE1 in wild-type increases the expression of the CBF regulon, resulting in improved frost resistance in transgenic plants (Chinnusamy et al., 2003).

9.3.3 Role of Calcium Toward Cold Stress

Calcium has a role in controlling the cold acclimation response because it is a significant second messenger in a signal transduction pathway induced at low temperature. Low temperatures trigger an upsurge in cytoplasmic calcium levels in both Arabidopsis and alfalfa, owing to the calcium influx from extracellular deposits. Calcium is essential for effective expression of at least certain cold-regulated genes in Arabidopsis, including the COR6.6 gene regulated by CRT/DRE and the KIN1 gene, in order to enhance the plant's freezing resistance. Calcium channel blockers (La3+) and calcium chelators (BAPTA) are also reported to inhibit cold-induced calcium influx in alfalfa, culminating in reduced expression of the cas15 gene (a cold-inducible gene) as well as the plant's potential to frost. Furthermore, they have discovered that the expression of the cas15 gene can be upregulated by treating cells with a calcium ionophore (A23187) at a temperature of 25°C that generates a substantial influx of calcium (Thomashow, 1999).

The Wcs120 gene encodes a protein that protects wheat against cold stress and represents a crucial function during cold stress. Multiple proteins interact with numerous components in nucleus extracts from non-acclimated plants. However, such proteins do not show any activity on DNA-binding induced in nuclear extracts from cold-acclimated plants (CA). Furthermore, the binding function of CA nuclear extracts is preserved by *in vitro* dephosphorylation with alkaline phosphatase and a phosphatase inhibitor (okadaic acid) significantly boosted the Wcs120 family protein's accumulation via *in vivo* (Vazquez-Tello et al., 1998). Wcs120 gene expression is expressed negatively by PP1 and/or PP2A (protein phosphatases). The research shows that CA nuclear extracts have substantially greater kinase activity that is dependent or independent of Ca2+. Thus, it is concluded that nuclear factors whose activity of binding is altered by a phosphorylation/dephosphorylation pathway may regulate the Wcs120 gene (Vazquez-Tello et al., 1998).

Another study of cold acclimation was reported by (Sangwan et al., 2001) on *Brassicanapus*. The promoter BN115 is activated by inducing a membrane structure change caused by cold temperature and further demands Ca2+ influx. By measuring the accumulation of BN115 transcripts, the BN115 promoter activity was assessed at both levels of transcription and translation. At 25°C, activation of BN115 by cADPR, Ca2+ ionophore, DMSO, and inhibitors of 2A and PP1 resulted in an enhanced tolerance to cold temperatures (Sangwan et al., 2001).

9.4 PLANT GENE REGULATION IN HEAVY STRESS

9.4.1 Metal Stress and MicroRNAs

Increased ROS production and cellular redox imbalance are caused when plants are exposed to metals at high concentrations. The cellular redox balance is disrupted; ROS levels rise, and oxidative damage or stress response signaling is triggered. Plants use miRNAs to control gene expression in response to metal toxicity via the mentioned modifications by Gielen et al. (2012). Those are: (1) excess metal complexation, (2) oxidative stress defense, and (3) signal transmission for managing diverse biological responses (Gielen et al., 2012).

9.4.2 Phytochelatin Synthesis

Phytochelatin synthesis is responsible for the detoxification of heavy metals in plants. The effect of Si in modulating gene expression is also credited with its

TABLE 9.1
The Effects of Different Types of Stress on Plants and Reported Genes to Regulate the Respective Abiotic Stress

Environmental Stress	Effect of Stress on Plants	Response Regulated Genes	Effect Of Plants After Stress Regulation	References
Drought stress	• Abnormal metabolism effecting normal growth • Plant mortality • Non-normal physiological processes • Decreased chlorophyll a and chlorophyll b • Closure of stomata • Reduces water use efficiency in plants • Rate of photosynthesis is reduced mainly by stomatal closure • Disturbed activity of various enzymes • Total leaf lipid content decreased	TaF3H CSD1 miR393 miR397b miR402 miR169 miR389a	• Maintains normal metabolism and normal growth • Restoring chlorophyll concentration • Normal stomata functioning • Activity of enzymes restored • Normal leaf lipid content • Normal photosynthesis and restoring water level in plants • Normal physiological processes	Fathi & Tari (2016) Mafakheri et al. (2010) Rizhsky et al. (2002) Farooq et al. (2012) Gigon et al. (2004)
Cold stress	• Poor germination • Stunted seedlings • Yellowing of leaves (chlorosis) • Reduced leaf expansion and wilting • Severe membrane damage • Changes in photosynthetic capacity • Affects the physiological, biochemical, molecular, growth, and developmental processes of plants • Flower abortion • Pollen and ovule infertility	CBF HOS1 SIZ1 BN115 CBF1 CBF2 CBF3 ICE1 CBF3 CRT/DRE COR6.6	• Proper germination and normal seedlings • Normal leaf structure with no wilting • No membrane damages • Normal photosynthesis • Normal physiological, biochemical, molecular, growth, and developmental processes of plants • Reduced reactive oxygen species (ROS) production • No senescence	Yadav (2010) Hajihashemi et al. (2018) Thakur et al. (2010) Karami-Moalem et al. (2018) Solanke & Sharma (2008) El-Mahdy et al. (2018)

	• Reactive oxygen species (ROS) production • Senescence occurs due to cold stress • Impairment of enzymatic activity • Misformation of protein synthesis	KIN1 cas15 Wcs120		
Heavy metal stress	• Senescence • Generation of reactive oxygen species • Peroxidation of lipids • Oxidation of protein • Decrease in chlorophyll • DNA damage • Inactivation of enzymes • Decrease in superoxide dismutase activity	miRNA OsHMA2 OsHMA3 OsLSi1 OsLSi2	• No senescence • No generation of reactive oxygen species • Normalized chlorophyll content • Reduced DNA damage • Normal activation of enzymes • Restored superoxide dismutase activity	Maksymiec (2007) Gill (2014) Gallego et al. (1996)

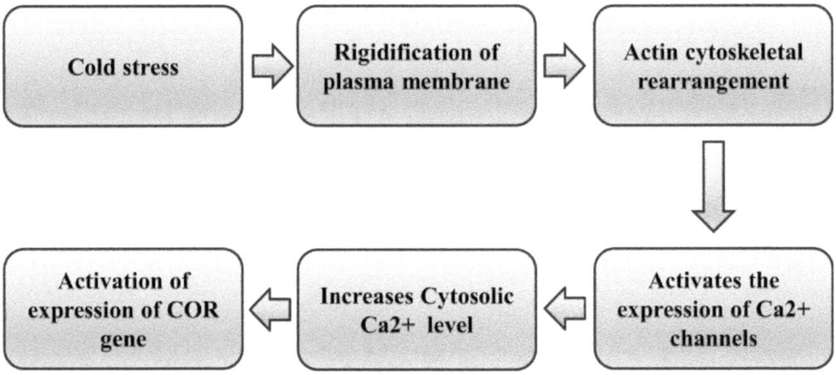

FIGURE 9.1 A flow chart of the response activity to cold acclimation in plants.

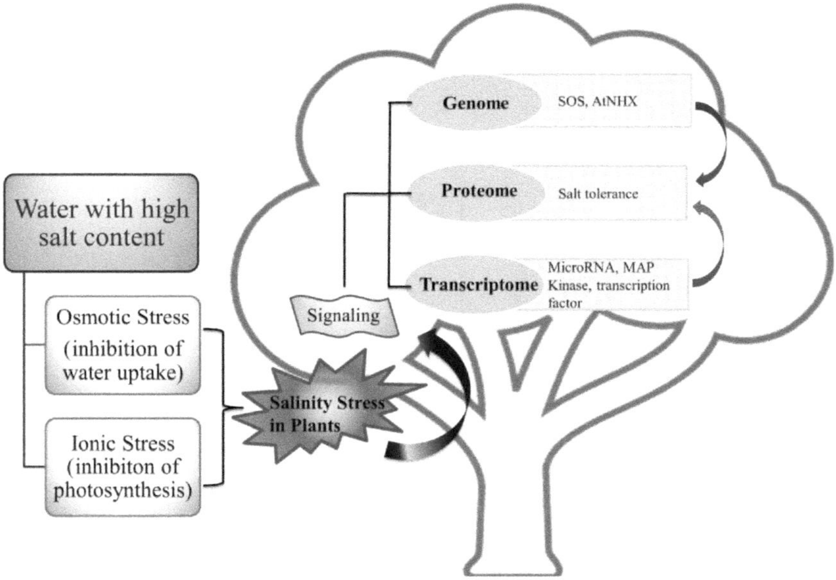

FIGURE 9.2 The response cascade induced by salinity stress by plants to trigger salt tolerance strategies.

ability to reduce metal toxicity. Si-mediated metal toxicity mitigation also contributes to the modulation of gene expression. Si supplementation triggered the genes that govern the synthesis of chelating agents and metallothioneins (MTs) in Arabidopsis growing under stress from copper. Similarly, incorporation of Si in rice resulted in a noteworthy upregulation of Si transporter encoding genes (OsLSi1 and OsLSi2) but substantial downregulation of genes encoding transporters of heavy metals (OsHMA2 and OsHMA3) (Bhat et al., 2019).

9.4.3 Plant Cell Antioxidant Network

Free radicals and ROS are precursors in a variety of plant metabolic reactions, such as lignin and cutin and lignin production, in non-stressed conditions. Plants have a complicated antioxidant defense mechanism that can be influenced by environmental factors to prevent oxidative and radical damage. Antioxidant enzymes such as catalases (CAT), super oxide dismutases (Fe-SOD, Cu/Zn-SOD, and Mn-SOD), and certain hydrogen peroxide reducing peroxidases are reported to be expressed in stress environments. Furthermore, other antioxidant enzymes, including glutathione peroxidases and lipid hydroperoxide reductase, are involved in the reductive detoxification of organic hydroperoxides produced by biomolecule peroxidation to prevent plants from metal toxicity (Dietz et al., 1999).

9.5 CONCLUSION

The abiotic stresses a plant generally experiences are prolonged drought, freeing temperatures, and toxicity due to the presence of heavy metals in the soil. The stress environment increases the production of ROS, which are primarily responsible for oxidative damage to plants. However, many plant species have adopted a robust antioxidant defense system to regulate, maintain, and exclude low intracellular ROS reservoirs. Plants, on the other hand, have an effective antioxidative defense mechanism that maintains, regulates, and excludes a low intracellular ROS pool (Pandey & Gautam, 2020). Plants that have been genetically changed have been produced and cultivated throughout the previous few decades, and they have been tested under a variety of stress environments. The techniques of genetic engineering are being utilized to produce plants with enhanced stress resilience; nevertheless, if full potential of this advancement is obtained, numerous obstacles must be addressed. There have been lots of practices in the last few years for using plants that are genetically modified to analyze responses to stress. This method may eventually result in the development of plants with increased stress tolerance. Many concerns, however, may be addressed if this strategy is used to its full potential. By regulating the genes involved in stress management, one can tackle problems that agricultural plants confront in the field. The transgenic plants research provides information on the involvement of various genes and antioxidant solute compounds that are expressed and required for cell survival under extreme circumstances and also provides their significance in crop yield (Wang et al., 2003). To survive extreme environments, a plant must adapt a specific defense mechanism and activate a response against stress stimuli (Maksymiec, 2007). All of these environmental stresses cause gene expression to be altered, as well as the production of novel proteins and the suppression of at least some regularly produced proteins, Furthermore, stress-induced proteins assist plants in making structural and biochemical modifications that aid the plant in coping with the stress circumstances that they are experiencing (Sachs & Ho, 1986). Plant responses to drought stress have been widely studied, revealing sophisticated regulatory gene networks, as well as physiological, cellular, and molecular responses, including the recognition of water deficiency signals. Plants detect water deficits at their roots and send a signal to their shoots, which causes abscisic acid (ABA) to be synthesized in the leaves (Takahashi et al., 2018). Drought causes the phytohormone abscisic acid (ABA) to be produced and/or mobilized, which is well known for its propensity to close stomata. Several of these elements cause signaling components in roots, shoots, and the environment to activate, all of which contribute to stomatal closure (Agurla et al., 2018).

REFERENCES

Agati, G., & Tattini, M. (2010). Multiple functional roles of flavonoids in photoprotection. *New Phytologist, 186*(4), 786–793. https://doi.org/10.1111/j.1469-8137.2010.03269.x

Agurla, S., Gahir, S., Munemasa, S., Murata, Y., & Raghavendra, A. S. (2018). Mechanism of stomatal closure in plants exposed to drought and cold stress. *Advances in Experimental Medicine and Biology, 1081*, 215–232. https://doi.org/10.1007/978-981-13-1244-1_12

Andrade, J. C., Terto, J., Silva, J. V., & Almeida, C. (2015). Expression profiles of sugarcane under drought conditions: Variation in gene regulation. *Genetics and Molecular Biology, 38*(4), 465–469. https://doi.org/10.1590/s1415-475738420140288

Bartels, D., & Sunkar, R. (2005). Drought and salt tolerance in plants. *Critical Reviews in Plant Sciences, 24*(1), 23–58.

Bhat, J. A., Shivaraj, S. M., Singh, P., Navadagi, D. B., Tripathi, D. K., Dash, P. K., Solanke, A. U., Sonah, H., & Deshmukh, R. (2019). Role of silicon in mitigation of heavy metal stresses in crop plants. *Plants (Basel), 8*(3). https://doi.org/10.3390/plants8030071

Chen, W. J., & Zhu, T. (2004). Networks of transcription factors with roles in environmental stress response. *Trends in Plant Science, 9*(12), 591–596. https://doi.org/10.1016/j.tplants.2004.10.007

Chinnusamy, V., Ohta, M., Kanrar, S., Lee, B. H., Hong, X., Agarwal, M., & Zhu, J. K. (2003). ICE1: A regulator of cold-induced transcriptome and freezing tolerance in Arabidopsis. *Genes & Development, 17*(8), 1043–1054. https://doi.org/10.1101/gad.1077503

Chinnusamy, V., Zhu, J., & Zhu, J. K. (2006). Gene regulation during cold acclimation in plants. *Physiologia Plantarum, 126*(1), 52–61.

Chinnusamy, V., Zhu, J., & Zhu, J. K. (2007). Cold stress regulation of gene expression in plants. *Trends in Plant Science, 12*(10), 444–451. https://doi.org/10.1016/j.tplants.2007.07.002

Covarrubias, A. A., & Reyes, J. L. (2010). Post-transcriptional gene regulation of salinity and drought responses by plant microRNAs. *Plant, Cell & Environment, 33*(4), 481–489. https://doi.org/10.1111/j.1365-3040.2009.02048.x

Daszkowska-Golec, A. (2016). The role of abscisic acid in drought stress: How ABA helps plants to cope with drought stress. In *Drought stress tolerance in plants* (Vol. 2, pp. 123–151). Springer.

Dietz, K.-J., Baier, M., & Krämer, U. (1999). Free radicals and reactive oxygen species as mediators of heavy metal toxicity in plants. In *Heavy metal stress in plants* (pp. 73–97). Springer.

Ding, Y., Tao, Y., & Zhu, C. (2013). Emerging roles of microRNAs in the mediation of drought stress response in plants. *Journal of Experimental Botany, 64*(11), 3077–3086. https://doi.org/10.1093/jxb/ert164

El-Mahdy, M., Youssef, M., & Eissa, M. (2018). Impact of *in vitro* cold stress on two banana genotypes based on physio-biochemical evaluation. *South African Journal of Botany, 119*, 219–225.

Fang, W., & Kao, C. H. (2000). Enhanced peroxidase activity in rice leaves in response to excess iron, copper and zinc. *Plant Science*, *158*(1–2), 71–76. https://doi.org/10.1016/s0168-9452(00)00307-1

Farooq, M., Hussain, M., Wahid, A., & Siddique, K. (2012). Drought stress in plants: An overview. *Plant Responses to Drought Stress*, 1–33.

Fathi, A., & Tari, D. B. (2016). Effect of drought stress and its mechanism in plants. *International Journal of Life Sciences*, *10*(1), 1–6.

Fowler, S., & Thomashow, M. F. (2002). Arabidopsis transcriptome profiling indicates that multiple regulatory pathways are activated during cold acclimation in addition to the CBF cold response pathway. *Plant Cell*, *14*(8), 1675–1690. https://doi.org/10.1105/tpc.003483

Gai, Z., Wang, Y., Ding, Y., Qian, W., Qiu, C., Xie, H., Sun, L., Jiang, Z., Ma, Q., Wang, L., & Ding, Z. (2020). Exogenous abscisic acid induces the lipid and flavonoid metabolism of tea plants under drought stress. *Scientific Reports*, *10*(1), 12275. https://doi.org/10.1038/s41598-020-69080-1

Gallego, S. M., Benavides, M. P., & Tomaro, M. L. (1996). Effect of heavy metal ion excess on sunflower leaves: Evidence for involvement of oxidative stress. *Plant Science*, *121*(2), 151–159.

Gielen, H., Remans, T., Vangronsveld, J., & Cuypers, A. (2012). MicroRNAs in metal stress: Specific roles or secondary responses? *International Journal of Molecular Sciences*, *13*(12), 15826–15847. https://doi.org/10.3390/ijms131215826

Gigon, A., Matos, A. R., Laffray, D., Zuily-Fodil, Y., & Pham-Thi, A. T. (2004). Effect of drought stress on lipid metabolism in the leaves of Arabidopsis thaliana (ecotype Columbia). *Annals of Botany*, *94*(3), 345–351. https://doi.org/10.1093/aob/mch150

Gill, M. (2014). Heavy metal stress in plants: A review. *International Journal of Advanced Research*, *2*(6), 1043–1055.

Hajihashemi, S., Noedoost, F., Geuns, J. M. C., Djalovic, I., & Siddique, K. H. M. (2018). Effect of cold stress on photosynthetic traits, carbohydrates, morphology, and anatomy in nine cultivars of stevia rebaudiana. *Frontiers in Plant Science*, *9*, 1430. https://doi.org/10.3389/fpls.2018.01430

Karami-Moalem, S., Maali-Amiri, R., & Kazemi-Shahandashti, S. S. (2018). Effect of cold stress on oxidative damage and mitochondrial respiratory properties in chickpea. *Plant Physiology and Biochemistry*, *122*, 31–39. https://doi.org/10.1016/j.plaphy.2017.11.011

Lee, H., Guo, Y., Ohta, M., Xiong, L., Stevenson, B., & Zhu, J. K. (2002). LOS2, a genetic locus required for cold-responsive gene transcription encodes a bi-functional enolase. *Embo Journal*, *21*(11), 2692–2702. https://doi.org/10.1093/emboj/21.11.2692

Ma, D., Sun, D., Wang, C., Li, Y., & Guo, T. (2014). Expression of flavonoid biosynthesis genes and accumulation of flavonoid in wheat leaves in response to drought stress. *Plant Physiology and Biochemistry*, *80*, 60–66. https://doi.org/10.1016/j.plaphy.2014.03.024

Mafakheri, A., Siosemardeh, A., Bahramnejad, B., Struik, P., & Sohrabi, Y. (2010). Effect of drought stress on yield, proline and chlorophyll contents in three chickpea cultivars. *Australian Journal of Crop Science*, *4*(8), 580–585.

Maksymiec, W. (2007). Signaling responses in plants to heavy metal stress. *Acta Physiologiae Plantarum*, *29*(3), 177–187.

Pandey, A. K., & Gautam, A. (2020). Stress responsive gene regulation in relation to hydrogen sulfide in plants under abiotic stress. *Physiologia Plantarum*, *168*(2), 511–525. https://doi.org/10.1111/ppl.13064

Parihar, P., Singh, S., Singh, R., Singh, V. P., & Prasad, S. M. (2015). Effect of salinity stress on plants and its tolerance strategies: A review. *Environmental Science and Pollution Research*, *22*(6), 4056–4075. https://doi.org/10.1007/s11356-014-3739-1

Park, S. Y., & Grabau, E. (2017). Bypassing miRNA-mediated gene regulation under drought stress: Alternative splicing affects CSD1 gene expression. *Plant Molecular Biology*, *95*(3), 243–252. https://doi.org/10.1007/s11103-017-0642-4

Rizhsky, L., Liang, H., & Mittler, R. (2002). The combined effect of drought stress and heat shock on gene expression in tobacco. *Plant Physiology*, *130*(3), 1143–1151. https://doi.org/10.1104/pp.006858

Sachs, M. M., & Ho, T.-H. D. (1986). Alteration of gene expression during environmental stress in plants. *Annual Review of Plant Physiology*, *37*(1), 363–376.

Sangwan, V., Foulds, I., Singh, J., & Dhindsa, R. S. (2001). Cold-activation of Brassica napus BN115 promoter is mediated by structural changes in membranes and cytoskeleton, and requires Ca2+ influx. *Plant Journal*, *27*(1), 1–12. https://doi.org/10.1046/j.1365-313x.2001.01052.x

Solanke, A. U., & Sharma, A. K. (2008). Signal transduction during cold stress in plants. *Physiology and Molecular Biology of Plants*, *14*(1–2), 69–79. https://doi.org/10.1007/s12298-008-0006-2

Takahashi, F., Kuromori, T., Sato, H., & Shinozaki, K. (2018). Regulatory gene networks in drought stress responses and resistance in plants. *Survival Strategies in Extreme Cold and Desiccation*, 189–214.

Thakur, P., Kumar, S., Malik, J. A., Berger, J. D., & Nayyar, H. (2010). Cold stress effects on reproductive development in grain crops: An overview. *Environmental and Experimental Botany*, *67*(3), 429–443.

Thomashow, M. F. (1999). Plant cold acclimation: Freezing tolerance genes and regulatory mechanisms. *Annual Review of Plant Physiology and Plant Molecular Biology*, *50*, 571–599. https://doi.org/10.1146/annurev.arplant.50.1.571

Vazquez-Tello, A., Ouellet, F., & Sarhan, F. (1998). Low temperature-stimulated phosphorylation regulates the binding of nuclear factors to the promoter of Wcs120, a cold-specific gene in wheat. *Molecular and General Genetics*, *257*(2), 157–166. https://doi.org/10.1007/s004380050635

Wang, W., Vinocur, B., & Altman, A. (2003). Plant responses to drought, salinity and extreme temperatures: towards genetic engineering for stress tolerance. *Planta*, *218*, 1–14.

Xiong, L., Ishitani, M., Lee, H., & Zhu, J. K. (2001). The Arabidopsis LOS5/ABA3 locus encodes a molybdenum cofactor sulfurase and modulates cold stress- and osmotic stress-responsive gene expression. *Plant Cell*, *13*(9), 2063–2083. https://doi.org/10.1105/tpc.010101

Yadav, S. K. (2010). Cold stress tolerance mechanisms in plants. A review. *Agronomy for Sustainable Development*, *30*(3), 515–527.

Yamaguchi-Shinozaki, K., & Shinozaki, K. (2006). Transcriptional regulatory networks in cellular responses and tolerance to dehydration and cold stresses. *Annual Review of Plant Biology*, *57*, 781–803. https://doi.org/10.1146/annurev.arplant.57.032905.105444

Yıldırım, K., & Kaya, Z. (2017). Gene regulation network behind drought escape, avoidance and tolerance strategies in black poplar (Populus nigra L.). *Plant Physiology and Biochemistry*, *115*, 183–199. https://doi.org/10.1016/j.plaphy.2017.03.020

Yoon, S. K., Park, E. J., Choi, Y. I., Bae, E. K., Kim, J. H., Park, S. Y., Kang, K. S., & Lee, H. (2014). Response to drought and salt stress in leaves of poplar (Populus alba × Populus glandulosa): Expression profiling by oligonucleotide microarray analysis. *Plant Physiology and Biochemistry*, *84*, 158–168. https://doi.org/10.1016/j.plaphy.2014.09.008

Zhu, J. K. (2002). Salt and drought stress signal transduction in plants. *Annual Review of Plant Biology*, *53*, 247–273. https://doi.org/10.1146/annurev.arplant.53.091401.143329

10 Role of Oxidative Stress-Responsive Thioredoxins in Biotic and Abiotic Stress Responses in Plants

Pragati Kumari, Rahul Thakur, Ashutosh Mani, Himanshu Kishore Prasad, and Saurabh Yadav

10.1 OXIDATIVE STRESS IN PLANTS

Plants are highly exposed to oxidative stress due to overproduction and accumulation of reactive oxygen species (ROS), a complex chemical and physiologic phenomenon triggered by different biotic and abiotic stresses such as high light, high or low temperature, salinity, drought, nutrient deficiency, and pathogen attack (Demidchik, 2015). ROS such as singlet oxygen ($^1O_2$), superoxide (O_2–), hydrogen peroxide (H_2O_2), and hydroxyl radical (HO.) are generated through the reduction of oxygen in chloroplasts, mitochondria, peroxisomes, and other cell sites during photosynthesis and respiration metabolic processes. Free radicals may harm the organism if they are not adequately inhibited (Nawkar et al., 2013). Unlike animals, plants cannot move from one place to another and are exposed to more biotic and abiotic stress (Vickers et al., 2009). Upon sensing, plants must deal with abiotic stress, such as heat, cold, light, UV, drought, and waterlogging, and biotic stress, due to bacteria, viruses, and fungi, with these threats simultaneously. To overcome these threats, plants have a well-defined coordinated defense mechanism to provide resistance toward all these abiotic and biotic stresses (Huang et al., 2019).

10.2 ROS HOMOEOSTASIS: ROLE OF REDOXINS

The molecules that remove the free radicals and the ROS are called antioxidants or free-radical scavengers. Antioxidants decrease or prevent the oxidation of biological molecules such as proteins, lipids, and DNA. Enzymatic and non-enzymatic biological processes may produce these antioxidants. Various antioxidants such as glutaredoxins, peroxiredoxins, thioredoxins, sulfiredoxins, and nucleoredoxins

DOI: 10.1201/9781003153481-10

are produced through enzymatic processes, collectively called redoxins (Bose et al., 2014).

10.3 THIOREDOXINS (TRXS)

Thioredoxins are small proteins with a redox-active disulfide bridge as the characteristic of an active site. The TRX oxidoreductase enzymes have a conserved active site with a cysteine-glycine-proline-cysteine (Cys-Gly-Pro-Cys) sequence pattern in the structural fold of all the TRX superfamily members. The molecular weight of TRX is ~12 kDa, and it is found in bacteria, fungi, plants, animals, etc. Thioredoxin proteins, such as glutaredoxins, are also involved in many plant–pathogen and pest interactions (Kumari, P., Gupta, A., & Yadav, S. 2021; Yadav et al., 2012). TRX superfamily members are involved in reversing diverse redox-based modifications (Spoel & Loake, 2011). In this chapter, we briefly discuss the roles of thioredoxin proteins in biotic and abiotic stresses.

10.4 BIOTIC STRESS

Whenever a plant is exposed to biotic stress by bacteria, viruses, herbivores, etc., the plant initiates a molecular signaling process to show a strong defense response, and many proteins, such as thionins, plant defensins, ribosome inactivating protein, cyclotides, pore-forming toxins, arcelins, canatoxin, urease enzymes, plant α-amylase inhibitors, protease inhibitors, and lectins are activated. Various studies have been reported that show that many genes upregulate and influence/modulate many genes/proteins to cope with biotic stresses upon the infection of different pathogens (Jaiswal et al., 2012). Through virus-induced gene silencing (VIGS), a thioredoxin-like 1 (TRXL1) gene was found to be involved in plant disease resistance and thermotolerance. TRXL1 is reduced, partly degraded via proteases and proteasome, and localized in chloroplasts during heat stress. Heat stress is an important stress that plants face at many stages of their life, and productivity is affected (Kumari et al., 2018, 2020). TRXL1 protein gets to interact with various proteins such as chaperonin (CPN60A), caseinolytic protease (CLPC1), and NADP-dependent malate dehydrogenase (NADPMDH). The protein chaperonin 60A (CPN60A) protects and guards TRXL1 against metabolic degradation, whereas CLPC1 protein degrades TRXL1 during heat stress. The activity of NADP-MDH gets regulated by TRXL1, and the malate level increases, thereby inhibiting the formation of superoxide radicals. Hence, it can be inferred that there is an antagonistic post-translational regulation of TRXL1 protein by CPN60A and CLPC1 and MDH regulation, helping in thermotolerance and plant disease resistance (Pant et al., 2020). In wheat, small GTP-binding protein TaRop10 was cloned and further characterized when *Puccinia striiformis* f. sp. *tritici (Pst)* infected wheat. It was observed that TaRop10 was differentially induced by abiotic stress and also by phytohormones. In response to *Pst* infection, plant thioredoxins TaTrxh9 and TaRop10 were found to play a key role in the negative regulation of defense signaling (Shi et al., 2021).

Rice is consumed globally, and its production is affected by the blast fungus *Magnaporthe oryzae*. This fungus attacks the rice's life stages and causes severe productivity loss (Zhang et al., 2016). Various studies were performed and the roles of oxidative stress-related proteins such as PR (pathogenesis related) proteins observed to delineate the molecular signaling processing from the biotic stress response.

10.5 ABIOTIC STRESS

Different types of abiotic stresses affect plants and their growth and development. So, studying the relationship between defense-responsive protein and abiotic stress seems important.

10.6 METAL STRESS

Various metals on earth have different roles, and how these metals affect plant development and productivity needs elaboration. So, the thioredoxin protein and its relation with the plant stress response is the main focus of this section.

There are many metals, including cadmium (Cd), that affects growth negatively. It is believed that Cd-impaired photosynthesis is one of the first responses that plants experience from Cd exposure (Rusinowski et al., 2019). Cd exposure reduces the plant's photosynthetic capacity, which is also associated with non-stomatal factors (Deng et al., 2014).

There are various proteins similar to thioredoins called thioredoxin-like proteins (TRX-like protein). CDSP32 (Trx CDSP32) is one such protein found in chloroplasts and regulated photosynthesis; it manages a plant's redox equilibrium under cadmium (Cd) stress. Exposure of cadmium to wild-type (WT) and overexpression lines of *Trx CDSP32* tobacco plants resulted in deciphering the role of Trx CDSP32 protein on photosynthetic machinery. It was observed that Trx CDSP32 protected the plants through cadmium-induced photoinhibition by two different photoprotective mechanisms (Zhang et al., 2021).

Nitrate is used by plants for growth and development and acts as fertilizer in agriculture to ensure the significant development of plants. It also acts as a signaling molecule and performs many physiological functions (Lee et al., 2021). Thioredoxins (TRX) plays a vital role in antioxidative defense, but little is known about tomato plants under nitrate stress. From tomatoes, the h-type TRX gene (*SlTrxh*) was isolated, and under nitrate stress, the transcript level was increased in roots and leaves. When excess nitrate was used in tobacco plants, beneficial effects were observed from overexpressing *SlTrxh*. Unlike wild-type, overexpressed plants exhibited higher germination rates, good root growth, and reduced ROS. But, when plants overexpressed Trxh, in which cysteine was mutated to serine at 54 positions, germination rate and root growth decreased after nitrate treatment. Transgenic tobacco plants overexpressing *SlTrxh* accumulated lower oxidative stress-related molecules and higher tolerance to methyl viologen (MV). It is well known that protein interacts with many other proteins during signaling and carries out important functions. When yeast two-hybrid was used in conjunction with other techniques for protein–protein interactions, peroxiredoxin (SlPrx) was isolated as an interacting partner of SlTrxh. Thus, thioredoxin (SlTrxh) helps maintain ROS homeostasis under nitrate stress by physically interacting with SlPrx (Zhai et al., 2022).

10.7 SALINITY STRESS

Salt stress limits the growth of plants, and this stress causes many other abiotic stresses. The salt concentration determines the stress level (Kumari, P., Gupta, A., Chandra, H., Singh, P., & Yadav, S. 2021). Overexpression and downregulation of transgenes were constructed to study the signaling under salt stress. If any of the downregulated transgenic lines performed well under salt stress, then some compensatory mechanisms are involved.

Transgenic Arabidopsis plants lacking TRXo1 responded in a similar manner to wild-type plants under salt stress of 100 mM. SOD, catalase, etc., enzymes may be replacing TRXo1 in maintaining ROS homoeostasis in the cell, which is important because TRXo1 is able to affect target proteins. Transgenic plants under salt stress provide insight into plant signaling under salt stress and how it correlates with ROS production (Calderón et al., 2018). Various enzymes – superoxide dismutase (SOD), catalase (CAT), peroxidases (POX), glutathione peroxidase (GPX), glutathione reductase (GR), glutathione S-transferases (GST), ascorbate peroxidase (APX), monodehydroascorbate reductase (MDHAR), and dehydroascorbate reductase (DHAR) – act with tolerance to oxidative stress, and one such enzyme, glutathione peroxidase, also called GPX and GPX-like (GPXL in Arabidopsis), helps in ROS quenching. This GPX-L is reduced by glutathione (GSH) and also TRXs (thioredoxin) (Dixon et al., 2009).

In plants, GPXs are reduced preferably by thioredoxins TRX instead of GSH as the reducing substrate (Li, 2016). In addition, plant GPXs contain cysteine (Cys) in their active sites. Mutation of AtGPXLs reduced TPOX activity under stress conditions, thereby showing a link with the TRX system. Research by Bela et al. (2018) showed a relationship between glutathione and thioredoxin.

Salt stress is also an important abiotic stress and causes a decrease in plant growth and yield, for instance in transgenic *Brassica napus* plants overexpressing AtTrx-*h*2 and AtTrx-*h*3 thioredoxin genes of *Arabidopsis thaliana*. Such transgenic plants provided salt tolerance on 50 mM NaCl, thus having more chlorophyll and fresh weight contents than non-transgenic plants. But, plants with the *AtTrx-h*3 gene have not provided salt tolerance, suggesting AtTrx-*h*2 is an important thioredoxin protein for crop improvement under salt stress (Ji et al., 2020).

10.8 HEAT STRESS

Temperature is crucial in plant growth and development, and plants are severely affected by increase and decrease in temperatures. Heat stress affects plant growth, photosynthesis, pollen development, and reproduction. Plant photosynthetic efficiency is significantly affected by the overproduction of reactive oxygen species due to the denaturation of heat shock proteins and alteration in enzyme activities (Ihsan et al., 2019). A thioredoxin-like 1 (TRXL1) gene was identified under virus-induced gene silencing (VIGS) screening, which is involved in disease resistance and also provides thermotolerance. During heat stress, TRXL1 protein is partly degraded via proteases and proteasome, and its chloroplast localization is changed. Also, the TRXL1 protein interacts with several proteins, including chaperonin, which protects it from degradation, and another protein, CLPC1, degrades TRXL1 under an increase in temperature stress (Pant et al., 2020).

10.9 COLD STRESS

Cold stress is another abiotic stress that involves multiple sets of molecules (Figure 10.1). Researchers have tried to find the molecular mechanism related to it and found the role of Trx-h2 under many abiotic stresses using mutant lines. The reductase function of Trx-h2 seems important for providing cold resistance in Arabidopsis (Table 10.1). Compared to Col-0 (wild-type) plants, Trx-h2 overexpression plants displayed strong cold-tolerant phenotypes (Park et al., 2021). Another protective

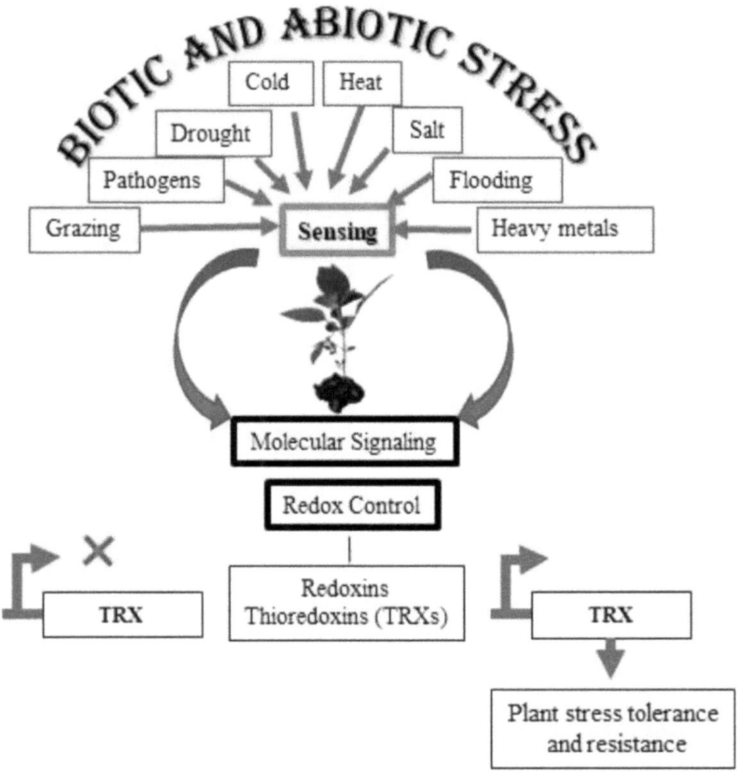

FIGURE 10.1 Representation of different stress and signaling by thioredoxin proteins, in plants.

function of thioredoxin was observed when an oxidoreductase enzyme TSV (temperature-sensitive virescent) interacted with OsTrxZ under low temperatures and had roles in chlorophyll synthesis (Sun et al., 2017).

10.10 CONCLUSION

Gene manipulation using transgenic technology has paved ways to modulate the level of TRX proteins, and this will be used to help in making stress-tolerant plants. For this, molecular signaling pathways related to oxidative stress and thioredoxins have been explored by researchers worldwide. Since TRX proteins act in coordinated manners, it is important to study how the modulation of TRXs modifies the expression level of other redox-related proteins. Thioredoxins and other such proteins of the TRX superfamily help in reducing oxidized proteins. So, molecular studies involving thioredoxins in biotic and abiotic stress in plants will throw further light on related areas, allowing plants to be engineered to achieve stress tolerance.

TABLE 10.1
Role of a Few Thioredoxins in Plants under Various Biotic and Abiotic Stress Conditions

Stress and Causal Agent		Plant Species	Stress Response	References
BIOTIC	*Pseudomonas syringae* pv. tomato	*Nicotiana benthamiana*	Antagonistic regulation of TRXL1 was observed at the post-translational level providing disease resistance	Pant et al. (2020)
	Puccinia striiformis f. sp. Tritici (Wheat stripe rust)	Wheat	In response to Pst infection defense signaling was negatively regulated	Shi et al. (2021)
	Magnaporthe oryzae (Rice blast)	*Oryza sativa*	MoAP1 transcriptionally regulates MoTrx2 which is important thioredoxin protein	Wang et al. (2017)
ABIOTIC	Cadmium stress	Tobacco	Cd-induced photoinhibition in tobacco leaves gets inhibited by Trx CDSP32	Zhang et al. (2021)
	Heat stress	*Nicotiana benthamiana*	During themotolerance there is antagonistic regulation of TRXL1 post-translationally	Pant et al. (2020)
	Nitrate stress	Tobacco	Under excess nitrate stress ROS homeostasis is regulated by thioredoxin (SlTrxh)	Zhai et al. (2022)

TABLE 10.1 ***(Continued)***
Role of a Few Thioredoxins in Plants under Various Biotic and Abiotic Stress Conditions

Stress and Causal Agent	Plant Species	Stress Response	References
Cold Stress	*Oryza sativa*	chloroplast development during cold stress is protected by an enzyme after protein–protein interaction with OsTrxZ	Sun et al. (2017)
Drought stress	*Arabidopsis thaliana*	Stressed Arabidopsis plants acclimatize by role of mitochondrial thioredoxin system	da Fonseca-Pereira et al. (2019)
Drought stress	*Arabidopsis thaliana*	During drought stress, new role of y-type TRXs as major antioxidants in plastids was displayed	Vanacker et al. 2018)
Drought stress	*Zea mays*	Trx-h overexpression was seen in BABA-primed leaves. Maize plants were able to handle more stressful conditions later	Shaw et al. (2017)
Cold stress	*Arabidopsis thaliana*	Cytosolic Trx-h2 protein was able to provide tolerance	Park et al. (2021)
Salt stress	*Brassica napus*	As compared to control, transgenic AtTrx-h2-overexpressing plants were able to withstand ROS stress caused by excess salt	Ji et al. (2020)
Salt and drought stress	*Phaseolus vulgaris*	Under combined salt and drought stress PvTrxh got upregulated under ROS	Boubakri et al. (2021)
Copper stress	*Pisum sativum*	Plant hormones modulates thioredoxin/ferredoxin systems and protect plants by less protein oxidative damage	Ben Massoud et al. (2018)
Salt and osmotic stress	*Arabidopsis thaliana*	AtGPXLs protected against oxidative stress	Bela et al. (2018)
Salt stress	*Arabidopsis thaliana*	Agronomical value of plants was increased after study on TRXo1	Calderón et al. (2018)

REFERENCES

Bela, K., Riyazuddin, R., Horváth, E., Hurton, Á., Gallé, Á., Takács, Z., Zsigmond, L., Szabados, L., Tari, I., & Csiszár, J. (2018). Comprehensive analysis of antioxidant mechanisms in Arabidopsis glutathione peroxidase-like mutants under salt- and osmotic stress reveals organ-specific significance of the AtGPXL's activities. *Environmental and Experimental Botany*, *150*, 127–140. https://doi.org/10.1016/j.envexpbot.2018.02.016

Ben Massoud, M., Sakouhi, L., Karmous, I., Zhu, Y., El Ferjani, E., Sheehan, D., & Chaoui, A. (2018). Protective role of exogenous phytohormones on redox status in pea seedlings under copper stress. *Journal of Plant Physiology*, *221*, 51–61. https://doi.org/10.1016/j.jplph.2017.11.014

Bose, J., Rodrigo-Moreno, A., & Shabala, S. (2014). ROS homeostasis in halophytes in the context of salinity stress tolerance. *Journal of Experimental Botany*, *65*(5), 1241–1257. https://doi.org/10.1093/jxb/ert430

Boubakri, H., Chihaoui, S.-A., Najjar, E., Gargouri, M., Barhoumi, F., & Jebara, M. (2021). Genome-wide analysis and expression profiling of H-type Trx family in Phaseolus vulgaris revealed distinctive isoforms associated with symbiotic N2-fixing performance and abiotic stress response. *Journal of Plant Physiology*, *260*, 153410. https://doi.org/10.1016/j.jplph.2021.153410

Calderón, A., Sánchez-Guerrero, A., Ortiz-Espín, A., Martínez-Alcalá, I., Camejo, D., Jiménez, A., & Sevilla, F. (2018). Lack of mitochondrial thioredoxin *o* 1 is compensated by antioxidant components under salinity in *Arabidopsis thaliana* plants. *Physiologia Plantarum*, *164*(3), 251–267. https://doi.org/10.1111/ppl.12708

da Fonseca-Pereira, P., Daloso, D. M., Gago, J., de Oliveira Silva, F. M., Condori-Apfata, J. A., Florez-Sarasa, I., Tohge, T., Reichheld, J.-P., Nunes-Nesi, A., Fernie, A. R., & Araújo, W. L. (2019). The mitochondrial thioredoxin system contributes to the metabolic responses under drought episodes in Arabidopsis. *Plant & Cell Physiology*, *60*(1), 213–229. https://doi.org/10.1093/pcp/pcy194

Demidchik, V. (2015). Mechanisms of oxidative stress in plants: From classical chemistry to cell biology. *Environmental and Experimental Botany*, *109*, 212–228. https://doi.org/10.1016/j.envexpbot.2014.06.021

Deng, G., Li, M., Li, H., Yin, L., & Li, W. (2014). Exposure to cadmium causes declines in growth and photosynthesis in the endangered aquatic fern (Ceratopteris pteridoides). *Aquatic Botany*, *112*, 23–32. https://doi.org/10.1016/j.aquabot.2013.07.003

Dixon, D. P., Hawkins, T., Hussey, P. J., & Edwards, R. (2009). Enzyme activities and subcellular localization of members of the Arabidopsis glutathione transferase superfamily. *Journal of Experimental Botany*, *60*(4), 1207–1218. https://doi.org/10.1093/jxb/ern365

Huang, H., Ullah, F., Zhou, D.-X., Yi, M., & Zhao, Y. (2019). Mechanisms of ROS regulation of plant development and stress responses. *Frontiers in Plant Science*, *10*. www.frontiersin.org/articles/10.3389/fpls.2019.00800

Ihsan, M. Z., Daur, I., Alghabari, F., Alzamanan, S., Rizwan, S., Ahmad, M., Waqas, M., & Shafqat, W. (2019). Heat stress and plant development: Role of sulphur metabolites and management strategies. *Acta Agriculturae Scandinavica, Section B – Soil & Plant Science*, *69*(4), 332–342. https://doi.org/10.1080/09064710.2019.1569715

Jaiswal, P., Cheruku, J. R., Kumar, K., Yadav, S., Singh, A., Kumari, P., Dube, S. C., Upadhyaya, K. C., & Verma, P. K. (2012). Differential transcript accumulation in chickpea during early phases of compatible interaction with a necrotrophic fungus Ascochyta rabiei. *Molecular Biology Reports*, *39*(4), 4635–4646. https://doi.org/10.1007/s11033-011-1255-7

Ji, M. G., Park, H. J., Cha, J.-Y., Kim, J. A., Shin, G.-I., Jeong, S. Y., Lee, E. S., Yun, D.-J., Lee, S. Y., & Kim, W.-Y. (2020). Expression of Arabidopsis thaliana Thioredoxin-h2 in Brassica napus enhances antioxidant defenses and improves salt tolerance. *Plant Physiology and Biochemistry*, *147*, 313–321. https://doi.org/10.1016/j.plaphy.2019.12.032

Kumari, P., Gupta, A., Chandra, H., Singh, P., & Yadav, S. (2021). Effects of salt stress on the morphology, anatomy, and gene expression of crop plants. In Pratibha Singh, Madhulika Singh, Rajiv Kumar Singh, Sheo Mohan Prasad (Eds.), *Physiology of salt stress in plants: Perception, signalling, omics and tolerance mechanism* (pp. 87–105). John Wiley & Sons Ltd.

Kumari, P., Gupta, A., & Yadav, S. (2021). Thioredoxins as molecular players in plants, pests, and pathogens. In I. K. Singh & A. Singh (Eds.), *Plant-pest interactions: From molecular mechanisms to chemical ecology* (pp. 107–125). Springer. https://doi.org/10.1007/978-981-15-2467-7_6

Kumari, P., Rastogi, A., & Yadav, S. (2020). Effects of heat stress and molecular mitigation approaches in orphan legume, Chickpea. *Molecular Biology Reports*, *47*, 4659–4670.

Kumari, P., Singh, S., & Yadav, S. (2018). Analysis of thermotolerance behaviour of five chickpea genotypes at early growth stages. *Brazilian Journal of Botany*, *41*(3), 551–565.

Lee, Y. J., Lee, W. J., Le, Q. T., Hong, S.-W., & Lee, H. (2021). Growth performance can be increased under high nitrate and high salt stress through enhanced nitrate reductase activity in arabidopsis anthocyanin over-producing mutant plants. *Frontiers in Plant Science*, *12*. www.frontiersin.org/articles/10.3389/fpls.2021.644455

Li, Y. (2016). *Redox regulation of salicylic acid synthesis in plant immunity*. https://era.ed.ac.uk/handle/1842/19578

Nawkar, G. M., Maibam, P., Park, J. H., Sahi, V. P., Lee, S. Y., & Kang, C. H. (2013). UV-Induced cell death in plants. *International Journal of Molecular Sciences*, *14*(1), 1608–1628. https://doi.org/10.3390/ijms14011608

Pant, B. D., Oh, S., Lee, H.-K., Nandety, R. S., & Mysore, K. S. (2020). Antagonistic regulation by CPN60A and CLPC1 of TRXL1 that regulates MDH activity leading to plant disease resistance and thermotolerance. *Cell Reports*, *33*(11), 108512. https://doi.org/10.1016/j.celrep.2020.108512

Park, J. H., Lee, E. S., Chae, H. B., Paeng, S. K., Wi, S. D., Bae, S. B., Thi Phan, K. A., & Lee, S. Y. (2021). Disulfide reductase activity of thioredoxin-h2 imparts cold tolerance in Arabidopsis. *Biochemical and Biophysical Research Communications*, *568*, 124–130. https://doi.org/10.1016/j.bbrc.2021.06.081

Rusinowski, S., Krzyżak, J., Sitko, K., Kalaji, H. M., Jensen, E., & Pogrzeba, M. (2019). Cultivation of C4 perennial energy grasses on heavy metal contaminated arable land: Impact on soil, biomass, and photosynthetic traits. *Environmental Pollution*, *250*, 300–311. https://doi.org/10.1016/j.envpol.2019.04.048

Shaw, A. K., K. Bhardwaj, P., Ghosh, S., Azahar, I., Adhikari, S., Adhikari, A., R. Sherpa, A., K., Saha, S., & Hossain, Z. (2017). Profiling of BABA-induced differentially expressed genes of Zea mays using suppression subtractive hybridization. *RSC Advances*, *7*(69), 43849–43865. https://doi.org/10.1039/C7RA06220F

Shi, B., Wang, J., Gao, H., Yang, Q., Wang, Y., Day, B., & Ma, Q. (2021). The small GTP-binding protein TaRop10 interacts with TaTrxh9 and functions as a negative regulator of wheat resistance against the stripe rust. *Plant Science*, *309*, 110937. https://doi.org/10.1016/j.plantsci.2021.110937

Spoel, S. H., & Loake, G. J. (2011). Redox-based protein modifications: The missing link in plant immune signalling. *Current Opinion in Plant Biology*, *14*(4), 358–364. https://doi.org/10.1016/j.pbi.2011.03.007

Sun, J., Zheng, T., Yu, J., Wu, T., Wang, X., Chen, G., Tian, Y., Zhang, H., Wang, Y., Terzaghi, W., Wang, C., & Wan, J. (2017). TSV, a putative plastidic oxidoreductase, protects rice chloroplasts from cold stress during development by interacting with plastidic thioredoxin Z. *New Phytologist*, *215*(1), 240–255. https://doi.org/10.1111/nph.14482

Vanacker, H., Guichard, M., Bohrer, A.-S., & Issakidis-Bourguet, E. (2018). Redox regulation of monodehydroascorbate reductase by thioredoxin y in plastids revealed in the context of water stress. *Antioxidants*, *7*(12), Article 12. https://doi.org/10.3390/antiox7120183

Vickers, C. E., Gershenzon, J., Lerdau, M. T., & Loreto, F. (2009). A unified mechanism of action for volatile isoprenoids in plant abiotic stress. *Nature Chemical Biology*, *5*(5), Article 5. https://doi.org/10.1038/nchembio.158

Wang, J., Yin, Z., Tang, W., Cai, X., Gao, C., Zhang, H., Zheng, X., Wang, P., & Zhang, Z. (2017). The thioredoxin MoTrx2 protein mediates reactive oxygen species (ROS) balance and controls pathogenicity as a target of the transcription factor MoAP1 in *Magnaporthe oryzae*: *M. oryzae* thioredoxin MoTrx2. *Molecular Plant Pathology*, *18*(9), 1199–1209. https://doi.org/10.1111/mpp.12484

Yadav, S., Kushwaha, H. R., Kumar, K., & Verma, P. K. (2012). Comparative structural modeling of a monothiol GRX from chickpea: Insight in iron – sulfur cluster assembly. *International Journal of Biological Macromolecules*, *51*(3), 266–273. https://doi.org/10.1016/j.ijbiomac.2012.05.014

Zhai, J., Qi, Q., Wang, M., Yan, J., Li, K., & Xu, H. (2022). Overexpression of tomato thioredoxin h (SlTrxh) enhances excess nitrate stress tolerance in transgenic tobacco interacting with SlPrx protein. *Plant Science: An International Journal of Experimental Plant Biology*, *315*, 111137. https://doi.org/10.1016/j.plantsci.2021.111137

Zhang, H., Liu, X., Zhang, H., Wang, Y., Li, T., Che, Y., Wang, J., Guo, D., Sun, G., & Li, X. (2021). Thioredoxin-like protein CDSP32 alleviates Cd-induced photosynthetic inhibition in tobacco leaves by regulating cyclic electron flow and excess energy dissipation. *Plant Physiology and Biochemistry*, *167*, 831–839. https://doi.org/10.1016/j.plaphy.2021.09.016

Zhang, H., Zheng, X., & Zhang, Z. (2016). The Magnaporthe grisea species complex and plant pathogenesis. *Molecular Plant Pathology*, *17*(6), 796–804. https://doi.org/10.1111/mpp.12342

Index

Page numbers in *italic* indicate a figure, and page numbers in **bold** indicate a table on the corresponding page.

A

B

C

D

E

F

G

L

M

N

O

P

Q

R

X

Z